225 SPECIES IN FULL COLOR

INSECTS

A GUIDE TO FAMILIAR AMERICAN INSECTS

a Golden Guide® from St. Martin's Press

by
HERBERT S. ZIM
and
CLARENCE COTTAM

Revised by
JONATHAN P. LATIMER
and
KAREN STRAY NOLTING
and
DAVID WAGNER,
UNIVERSITY OF CONNECTICUT

Illustrated by
JAMES GORDON IRVING and SUSAN SIMON

St. Martin's Press New York

FOREWORD

Many individuals and institutions have helped make this guide a success. The authors, Herbert S. Zim, who conceived Golden Guides, and Clarence Cottam, created a work of lasting importance. The artist, James Gordon Irving, and his wife, Grace Crowe Irving, made a major contribution. Specialists consulted for the first edition included Robert T. Mitchell, William D. Field, Edward A. Chapin, William H. Anderson, Austin H. Clark, George B. Vogt, Reece I. Sailer, Hahn W. Capps, O. L. Cartwright, Paul W. Oman, Ashley B. Gurney, Barnard D. Burks, Karl V. Krombein, Ross H. Arnett, Jr., Marion R. Smith, Alan Stone, John G. Franclemont, Arthur B. Gahan, Curtis W. Sabrosky, Grace B. Glance, C. F. W. Muesebeck, and Jean Laffoon. Allen M. Young, Curator and Head, Invertebrate Zoology, Milwaukee Public Museum, prepared an earlier revision.

Our thanks go to the specialists consulted for this revision: Derek Sikes, John Colby, Jon Gelhaus, Charles Henry, Chris Maier, Piotr Naskrecki, Jane O'Donnell, Carl Schaefer, Mike Sharkey, Mike Wall, John Weaver, and John Wenzel. Special thanks also go to Julie Henry for her enormous contribution of time and talent. We hope this guide will continue to help readers of all ages recognize and appreciate the insects around us.

J. P. L.
K. S. N.

 Printed in China. St. Martin's Press, 175 Fifth Avenue, New York, N.Y. 10010. www.stmartins. com

ISBN 978-0-312-53035-8

USING THIS BOOK

By dealing with common, important, and showy insects, this book will help you begin a fascinating study. To identify an insect, turn to the key to the insect groups (orders) on pages 4 and 5. Insects are grouped by orders; each order contains insects with certain characteristics in common. Compare your insect with the illustrated example used to represent an order and look for similar features. Carefully read the descriptions to find out which fits your specimen best. When you think you know to what order your insect belongs, turn to the pages given. There you will find more insects from that order to compare with yours. A strong magnifying lens can be very helpful.

Insects in this book are usually shown on their food plants. Immature forms often appear with the adult. If you cannot identify an immature insect, try to rear it to maturity.

In advanced study, the Latin scientific names of species are used for greater precision in designation. Scientific names of species illustrated in this book are given on pages 155–157.

On plates, approximate lengths are given in inches ("w." indicates wingspread).

Range maps show occurrence of species within the United States, just over the Mexican border, and about 200 miles northward into Canada. Since distribution of many species is not well known and may change over time, ranges given are only approximate. Where ranges of two or more insects appear on one map, each has a different color or line pattern, as in the sample here. A magenta tint over a line pattern indicates greater abundance (as on page 20). Each caption is on or next to the color to which it refers.

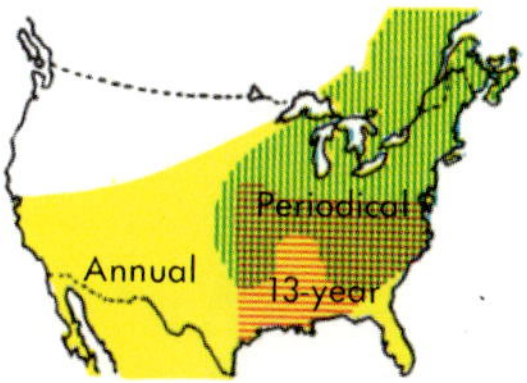

KEY TO MAJOR INSECT GROUPS

MAYFLIES (Ephemeroptera) **AND STONE-FLIES** (Plecoptera), page 17. Both with two pairs of transparent, many-veined wings. In mayflies, hindwings smaller; in stoneflies, larger. Mayflies have long, pronged "tails." Immatures aquatic.

DRAGONFLIES AND DAMSELFLIES (Odonata), pages 18–19. Medium-sized to large insects with two pairs of long, equal-sized wings. Body long and slender. Antennae short. Immature stages aquatic. Development in three stages.

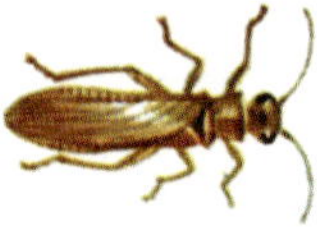

GRASSHOPPERS, ROACHES, AND THEIR KIN (Orthoptera), pages 20–31. Medium-sized to large insects. Live on land. Forewings leathery. Hindwings folded fan-like (some have no wings). Chewing mouth-parts. Development gradual.

TERMITES (Isoptera), pages 32–33. Ant-like insects, small and soft-bodied; workers mostly white. Some have four long wings. Live in colonies. Chewing mouth-parts. Development gradual.

EARWIGS (Dermaptera), page 34. Medium-sized insects with pincer-like tail. Forewings very sharp and squared off. Segmented antennae. Development gradual.

LEAFHOPPERS, APHIDS, SCALE INSECTS, AND CICADAS (Homoptera), pages 35–43. Small to medium-sized insects, most with two pairs of similar wings held sloping at sides of body. Jointed beak for sucking attached to base of head. Land insects. Some scale-like. Development gradual.

TRUE BUGS (Hemiptera), pages 44–51. Range from small to large. Two pairs of wings, with forewings partly thickened. Jointed beak for sucking arises from front of head. Development gradual.

LICE (Anoplura), page 52. Tiny, flattened, wingless insects with piercing and sucking mouthparts. Body flattened. Legs with claws for clinging to warm-blooded animals. Development gradual.

NERVE-WINGED INSECTS (Neuroptera), pages 53–55. Small to large insects. The two pairs of wings, usually equal in size, are netted with veins. Chewing mouth-parts. Long antennae. Four stages of development: egg, larva, pupa, and adult.

BEETLES (Coleoptera), pages 56–82. Forewings modified to thickened covers. Hindwings (sometimes absent) thin, folded. Size from tiny to large. Chewing mouth-parts. Antennae usually short. Some aquatic. All have four life stages.

CADDISFLIES (Trichoptera), page 83. Small to medium-sized insects. Most larvae live in fresh water. Some build ornamented cases. Adults with two pairs of wings with long, silky hairs and long forward-held antennae. Mouth-parts reduced. Development in four stages.

BUTTERFLIES AND MOTHS (Lepidoptera), pages 84–127. Small to large insects with two pairs of scaly wings. Sucking mouth-parts, sometimes reduced. Antennae knob-like, a single filament, or feathery. Development in four stages.

SCORPIONFLIES (Mecoptera), page 128. Medium-sized insects with two pairs of slender, generally spotted wings. Legs and antennae long. Beak-like, chewing mouth-parts. Larvae live in soil. Development in four stages.

FLIES AND THEIR KIN (Diptera), pages 129–135. Tiny to medium-sized insects, with sucking mouth-parts and two pairs of wings. Antennae small, eyes large. Second pair of wings highly reduced serving as a balancing organ. Development in four stages.

BEES, WASPS, AND ANTS (Hymenoptera), pages 136–149. Tiny to large insects; many social or colonial. Two pairs of thin, transparent wings. Hindwings smaller. Mouth-parts for chewing or sucking. Females may have "stingers." Development in four stages.

SEEING INSECTS

Insects have been on this earth for nearly 400 million years and are found nearly everywhere, even in the Antarctic. More kinds of insects are known than all other animals visible to the naked eye. A few insects have been called man's worst enemy, but we would be hard put to exist without them. Insects play many important roles in the environment. They are an important food source for many animals, extraordinary pollinators, and play a critical role in recycling. They are also gems of natural beauty, zoological mysteries, and a constant source of interest.

WHAT INSECTS ARE Insects are related to crabs and lobsters. Like these sea animals, they possess a kind of skeleton on the *outside* of their bodies. The body itself is composed of three divisions: head, thorax, and abdomen. The thorax has three segments, each with a pair of jointed legs; so an insect normally has six legs. Most insects also have two pairs of wings attached to the thorax, but some have only one pair, and a few have none at all. Insects usually have three sets of mouthparts, two kinds of eyes—simple and compound—and one pair of antennae.

PARTS OF AN INSECT

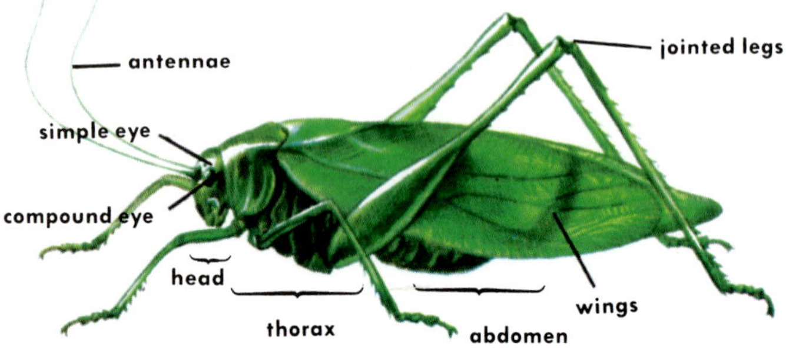

This describes typical insects, but many common ones are not typical. The thorax and abdomen may appear to run together. Immature stages (larvae) of many insects are worm-like, though their six true legs and perhaps some extra false ones may be counted. Immature insects are often difficult to identify. It is also hard to tell the sex of some insects. In some groups males are larger or have larger antennae or different markings. Females are sometimes marked by a pointed ovipositor for laying eggs extending from the base of the abdomen.

INSECT RELATIVES A number of animals are often confused with insects. Spiders, ticks, and mites have only two body divisions and four pairs of legs. They have no antennae. Other insect-like animals have the head and thorax joined like the spiders. Crustaceans have at least five pairs of legs and two pairs of antennae. Most live in water (crabs, lobsters, shrimps), but the sowbug is a land dweller. Centipedes and millipedes have many segments to their bodies with one pair of legs (centipedes) or two pairs (millipedes) on each. Centipedes have a pair of long antennae; millipedes have a short pair.

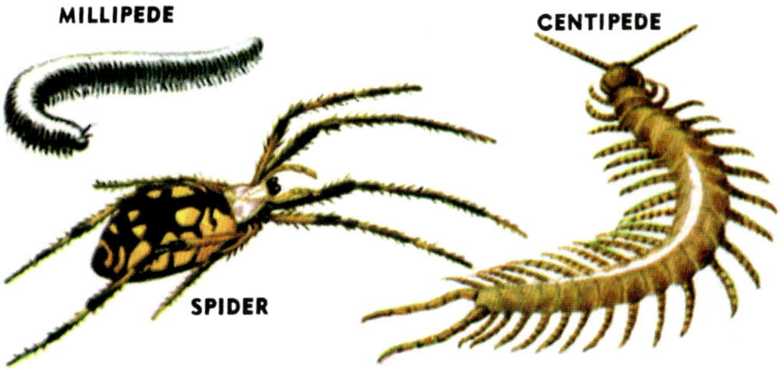

NUMBER OF INSECTS The insect group (Class Insecta) is by far the largest group of animals in the world. Over a million species have been identified, but one authority estimates this may be only 3 percent of the insects yet to be discovered. The class is divided into over 30 orders. One order encompasses the butterflies and moths; one, the termites; another, the beetles. The beetles alone include some 280,000 described species. There are more kinds of beetles than kinds of all other animals known, outside the insects. Butterflies and moths total over 146,000 species. Bees, wasps, and ants number 115,000; true bugs, 65,000 or more. The student of insect life need never run out of material. Over 15,000 species have been found around New York City alone. You can find a thousand species in your vicinity if you look for small insects as well as large, showy ones.

INSECTS AND PEOPLE Whether certain insects are considered helpful or harmful to people depends as much on us as it does on insects. Our ways of farming and raising animals have provided some insects which might otherwise be rare with conditions enabling them to multiply a thousandfold. Less than 1 percent of insects are considered harmful, but these destroy 10 percent of our crops, causing a loss of billions of dollars annually. Some insects are parasitic on other animals, and some carry diseases.

On the other hand, this would be a sorry world without insects. We would have no apples, grapes, or clover, much less cotton, and fewer oranges and garden vegetables, for these and many other plants depend on insects to pollinate their flowers. And there would be no honey, of course. Some insects aid the process of recycling, a process that is essential to life. Some insects help control others, and all help maintain the vitality of ecosystems.

INSECTS IN NATURE In the broad view, insects play an important natural role, not only in ways that benefit humans but in ways that make our rich plant life and wildlife possible. They are food for many kinds of mammals, birds, amphibians, and fish. Many of our songbirds depend almost entirely on an insect diet. Every fisherman knows how freshwater game fish go after insects. Keep this broad view in mind when people start talking about widespread insect control—something that may become possible with newer chemicals and genetic engineering. Pesticide usage on a large scale is devastating to other species, many of which may be rare. Widespread insect control may also have unexpected consequences, creating more problems than it solves.

CONTROL OF INSECTS There are ways to supplement the natural control of insects by birds and other animals. We encourage those harmless insects that prey on harmful kinds. We can exclude insects with screens, discourage them with repellents, trap them, or poison them. Since there are so many kinds of insects that live and feed in so many different ways, there is no single best method to get rid of them. Yet, with concerted effort some dangerous insects have been wiped out over fairly large areas. A unique example of this was the complete destruction of the Mediterranean fruit fly in Florida, which threatened the citrus crop in 20 counties.

If you have an important insect problem, consult your County Agricultural Extension Agent. Often entomologists (insect specialists) at universities or museums can help, or you can turn to the U.S. Department of Agriculture's Insect Identification Laboratory, Agricultural Research Center West, Beltsville, MD 20705, where experts work on nearly every phase of insect life and control. Many of these organizations also have excellent Web sites.

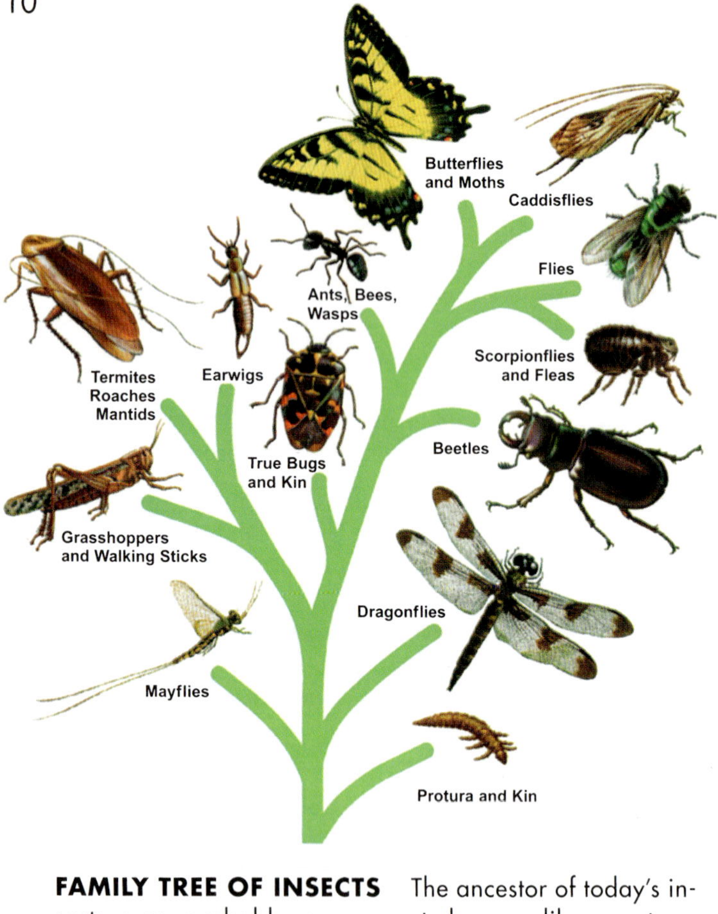

FAMILY TREE OF INSECTS The ancestor of today's insects was probably a segmented worm-like creature. Roaches and many other insects have been around for as long as 200 million years. Today there are over 30 orders of insects (depending on the classification), including over one million described species. Most of the 12,000 kinds of fossil insects identified so far are similar to living species. The illustration above shows some of the basic relationships among insects.

SILVERFISH—primitive insect with direct, gradual development

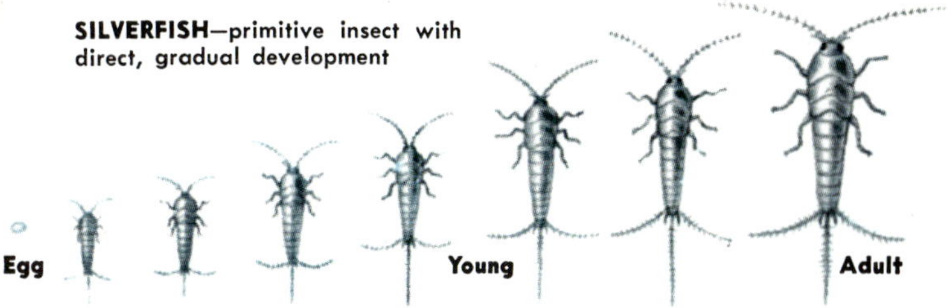

CHINCH BUG—more advanced insect, with incomplete metamorphosis or development in three stages

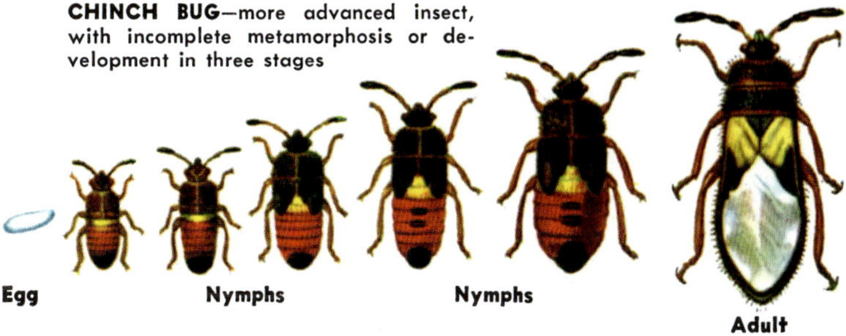

HOUSE FLY—advanced, modern insect, with complete metamorphosis or development in four growth stages

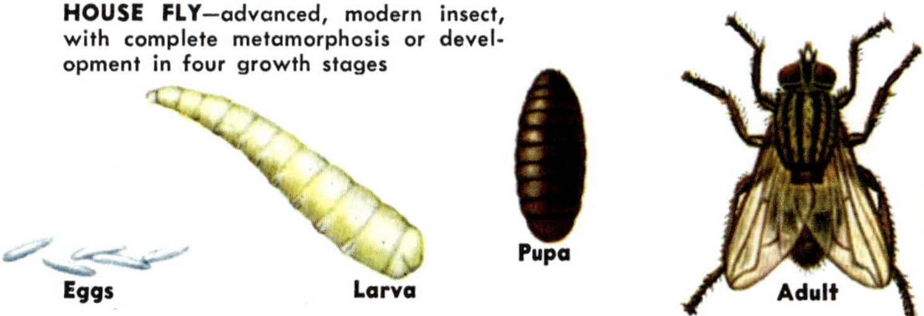

Insects follow different developmental patterns. In the simplest, the newly hatched insect is like a miniature adult. It grows and molts (sheds its skin) till it reaches adult size. In incomplete metamorphosis an immature nymph hatches, grows, and molts. It develops wings only in the last molt. Complete metamorphosis involves (1) egg, (2) larva, (3) pupa or resting stage, and (4) adult. In all three of the developmental patterns reproductive structures are only present or functional following the final molt.

INTERIOR OF A GRASSHOPPER (Side View)

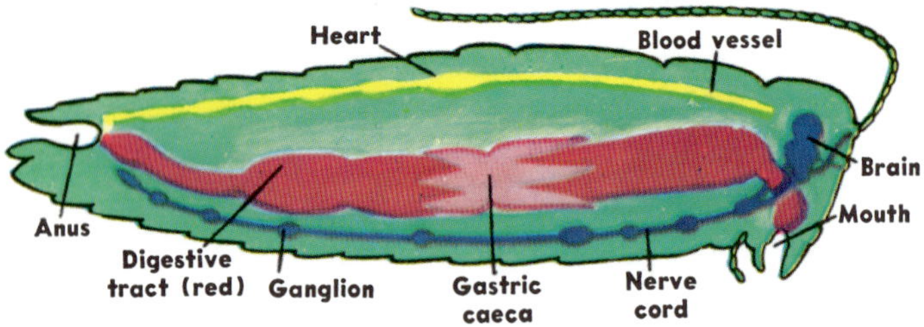

INSECT STRUCTURE is marked by three body divisions (page 6), six jointed legs, one pair of antennae, and usually one or two pairs of wings. The outer covering, or exoskeleton, is often hardened. Mouth-parts consist of three separate pairs of structures. Internally, insects have a digestive tract and auxiliary digestive organs. Breathing is done by air tubes spreading internally from openings called spiracles. Circulation is open with many of the internal organs being bathed in the green to yellow fluid called hemalymph. Respiration, digestion, and circulation are shown in the longitudinal section (above) and cross-section (below left) of the grasshopper, a typical insect. The nervous system (below) shows the brain, which is divided into three parts. Ganglions serve as nerve centers for nearby parts of the body.

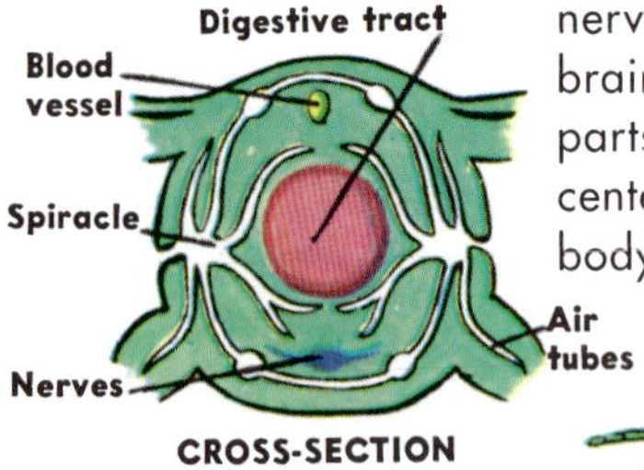

CROSS-SECTION

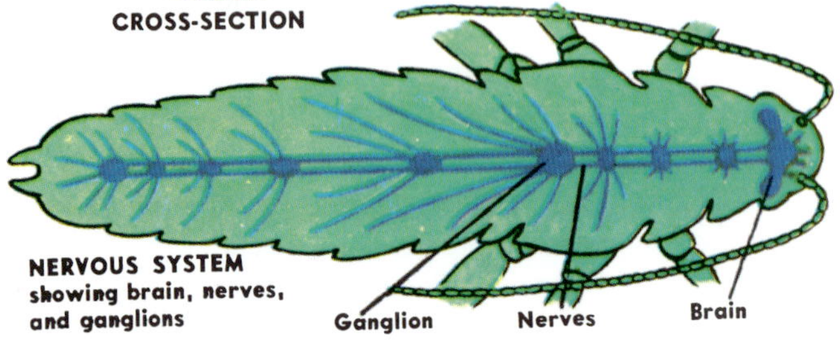

NERVOUS SYSTEM
showing brain, nerves, and ganglions

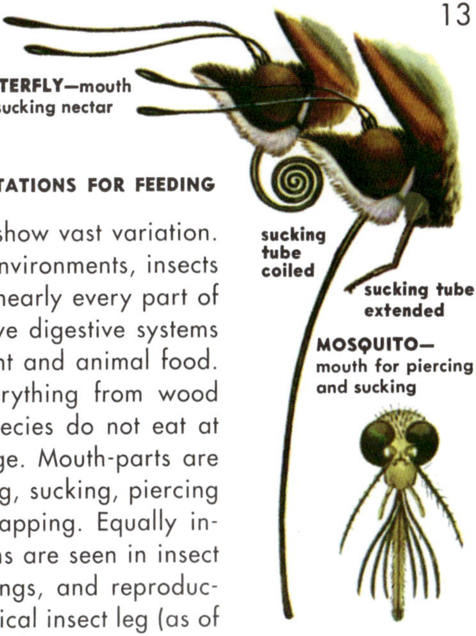

ADAPTATIONS FOR FEEDING

Insect structures show vast variation. Adapted to many environments, insects live successfully in nearly every part of the world. They have digestive systems for all kinds of plant and animal food. They thrive on everything from wood to blood. A few species do not eat at all in the adult stage. Mouth-parts are adapted for chewing, sucking, piercing and sucking, and lapping. Equally interesting adaptations are seen in insect wings, body coverings, and reproductive organs. The typical insect leg (as of a grasshopper) has five parts. Grasshopper hind legs are specialized for jumping. The housefly has pads that enable it to walk up windows or across ceilings. In honeybees, the hind leg has special hairs that carry pollen. In many insects the foreleg is used to groom and clean the antennae and body. Insect structures are fascinating to study under a lens.

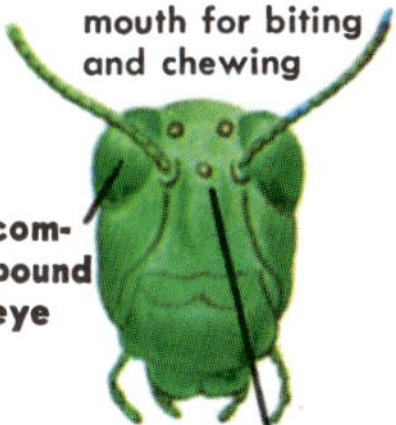

ADAPTATIONS OF FEET

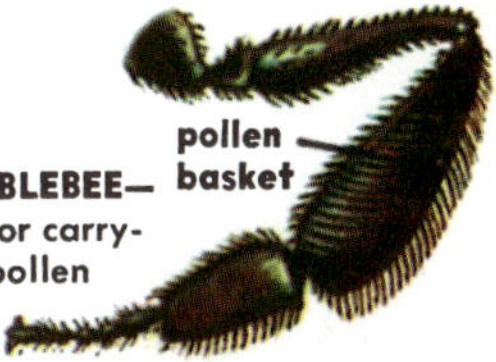

STUDYING AND COLLECTING INSECTS

Our knowledge of many insects is still so incomplete that a serious amateur can look forward to making important and lasting discoveries.

WHERE TO LOOK Practically everywhere: in fields, gardens, woods, roadsides, beaches, and swamps; under stones, rotted logs, and leaves. Look in flowers, on grass, on animals, too. You'll find insects in the air, in or on water, on and in the ground. Sunny areas often yield greatest insect diversity.

WHEN TO LOOK Insects are most common in late spring and summer, but experienced collectors can find them in all seasons. Many groups of insects are more common at night; remember that when collecting. In winter, concentrate on protected spots, such as under stones or bark, and in water. Watch for insects in egg cases or in their resting stage (pupae).

WHAT TO DO Studying insects is not confined to catching them and mounting them in collections. Raising and housing live insects to study their habits is exciting. Anyone can have an insect zoo in old glass jars. Collect immature insects (larvae), provide them with proper food, and watch

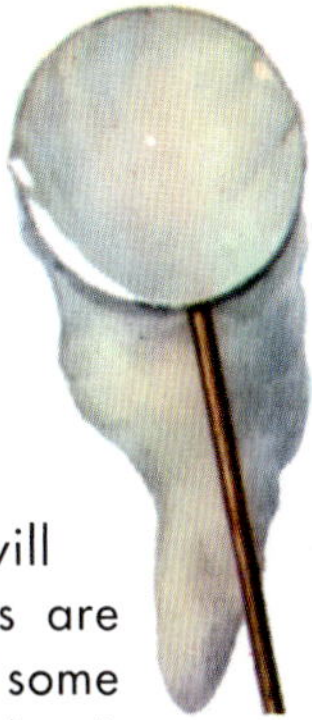

them grow. Watch caterpillars shed their skins, spin a cocoon or form a chrysalis, and emerge as a moth or a butterfly. See worm-like larvae become flies or beetles. Raise a colony of ants, bees, or termites. You will learn more from live insects than from dead specimens.

Whatever you do with insects, you will need some understanding of what insects are and how they live. Use this book, then read some of the other books suggested. Most important, go out and look at insects. Catch them if you wish, but watch them first. See how they move, how they feed, and what they do.

COLLECTING INSECTS An insect collection can be valuable for study or reference—if it is used. Using a collection means more than making a collection, though this step must come first. Fortunately, beginners can collect insects with simple, low-cost equipment. Not all insects are chased with a net. Hang an old bedsheet out at night with a light in front of it. Similar traps are described in reference books. Try the easiest methods and places first—at your window screens or near a large neon sign. Just gather the insect harvest there.

EQUIPMENT Any large, wide-mouthed jar will serve for confining and raising insects. Tie some gauze or netting over the top. A light net with a long handle is good for catching insects on the wing. A heavier

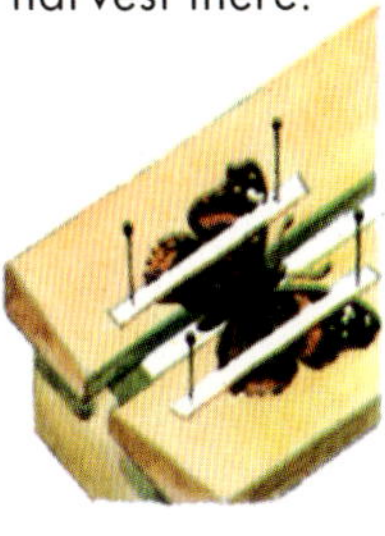

net is better for "sweeping" through the grass or for catching water insects.

Flat boxes with a layer of heavy corrugated cardboard on the bottom to hold pins are fine for storing specimens. Use a mothball to deter insect pests. Purchase and use insect pins; ordinary ones are too heavy and prone to rusting. Learn the tricks that make mounting neat and attractive. A book for records is essential; so are labels. Later you may want spreading boards, pinning blocks, and other accessories. Collecting and preserving specimens requires patience and skill. Read first and then practice with any insects you may find in your own yard. Skill will come with experience. A well-pinned, fully labeled specimen may retain its value for hundreds of years.

FIELD AND LIFE-HISTORY STUDIES may prove more interesting and exciting than collecting. Instead of learning a little about many insects, learn a lot about a few. Field studies can involve unusual problems on which there is little or no scientific information. How do ants recognize one another? How does temperature affect the flight of butterflies? How much does a caterpillar eat? Can beetles recognize color? Such problems can be investigated in your own yard if you are interested. Many insects are known only in the adult form; few facts are known about the rest of their life cycles. Constant observation of wild specimens, or detailed study of captive ones reared under natural conditions, will yield new and interesting facts.

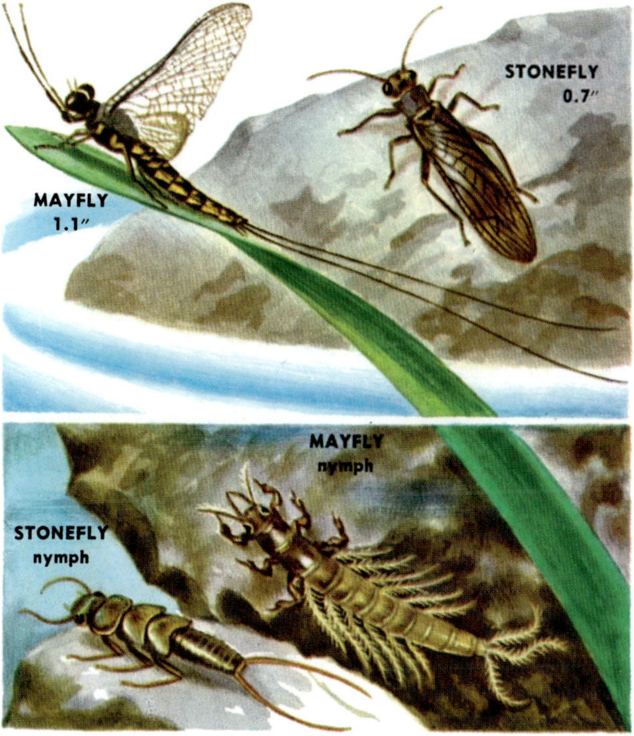

MAYFLIES AND STONEFLIES are unrelated yet similarly adapted to aquatic environments. The 600 or so species of mayflies have transparent, veined wings and a long, forked tail. The adults are short lived, sometimes living only 12 to 24 hours. The stoneflies include some 500 species. The nymphs, like those of mayflies, live in water and are important food of fresh-water fish. Some nymphs of both groups take several years to reach the adult stage. Adult stoneflies have transparent wings, though they are not strong fliers.

DRAGONFLIES AND DAMSELFLIES are often seen near rivers, streams, bogs, ponds, and moist meadows, but some species dwell in forests. Dragonflies, also known as darning needles or stingers, hunt small insects like mosquitoes, which they eat on the wing. It is believed that a single large dragonfly eats dozens of mosquitoes every evening. Dragonflies rest with wings outstretched. The more delicate damselfly rests with wings folded. Both lay eggs in water; the nymphs develop there, feeding on other aquatic insects. They leave the water after several growing stages; the skin splits and the adult emerges.

TEN-SPOT
DRAGONFLY
2.0″
cast
skin
arrowhead
BLACKWING
DAMSELFLY 1.3″

WALKINGSTICKS are large, usually wingless insects with legs all about the same length and shape, distinguishing them from the mantises (pp. 30–31). Walkingsticks live and feed on leaves of oak, locust, cherry, walnut, and other woody plants, occasionally causing damage. The female's 100 or so eggs are dropped singly to the ground to hatch the following spring. As the young grow, they molt or shed their skin five or six times; otherwise they are similar to adults. Males are smaller than females.

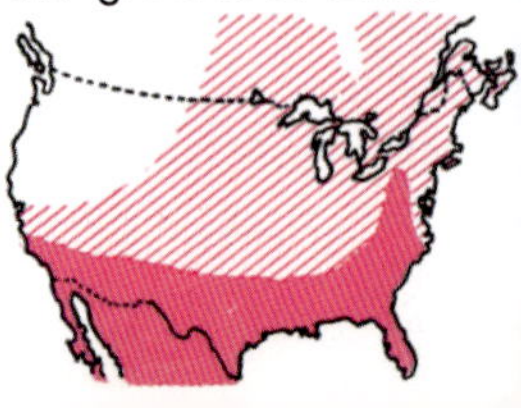

MOLE AND CAMEL CRICKETS These nocturnal crickets live under rocks in moist places, or mostly underground. The large mole cricket burrows near the surface, eating young roots and killing seedlings. In the South, it may destroy peanuts, strawberries, and other garden crops. The pale brown, spotted, wingless camel cricket is identified by its high, arched back. Though scavengers, they often become nuisances around greenhouses.

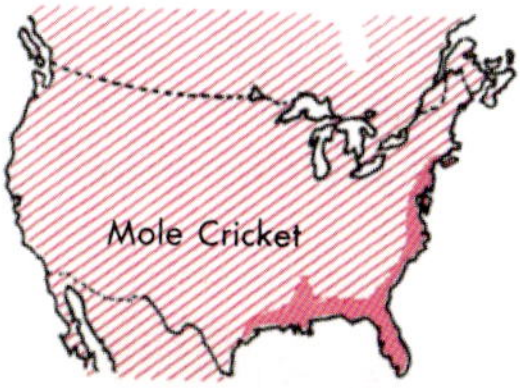

GRASSHOPPERS AND LOCUSTS All have short antennae and large hearing organs on the sides of the abdominal base. When disturbed, some regurgitate a noxious droplet ("tobacco juice") that discourages many would-be predators. Most are good fliers, though some are wingless. Locusts are merely grasshoppers that migrate, with two distinct phases: solitary and migratory. The latter are more brightly colored and highly gregarious. Females lay 20 to 100 eggs in the ground or in rotted wood. See p. 24 for the life history of one species. Nymphs mature in 2 to 3 months.

LUBBER GRASSHOPPER
2.5″
MIGRATORY
GRASSHOPPER 1.3″
sun-flower
wheat

RED-LEGGED GRASSHOPPER

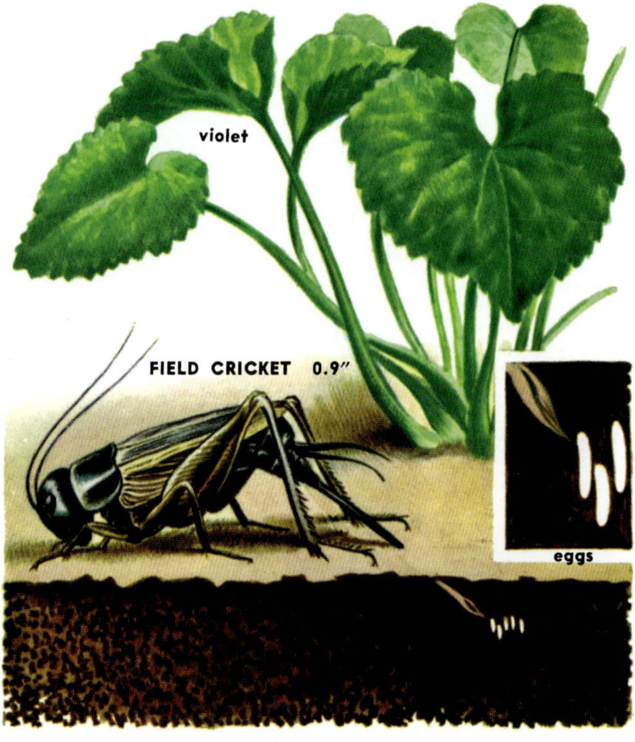

FIELD CRICKETS These common, large-headed black or brown crickets are largely nocturnal. Their shrill musical song is made by rubbing the edges of their forewings together. Though principally vegetarians, they occasionally eat other insects. Eggs are laid in the ground in fall. The young nymphs emerge in spring and develop their adult wings in several stages by late summer. Species that are similar in appearance may be distinguished by their song.

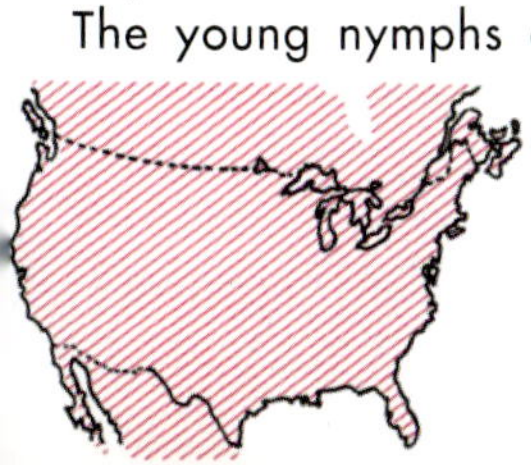

KATYDIDS The male of the true katydid makes the persistent "katydid" call of summer nights. At the base of the outer wings or wing-covers of the males are rasps and ridges which, when rubbed like a fiddle and bow, produce the calls of different species. Katydids hear by "ears" on the upper part of their front legs. Some katydids are tree dwellers, feeding on leaves of cherry, oak, maple, apple, and other woody species. Others live in weeds and shrubs. Most are green, with thin, leaf-like wing-covers,

and so have the advantage of protective coloration. However, some species are brown or even pink. All have long antennae. Females, recognized by the long ovipositor, usually place 100–150 eggs on leaves or twigs early in fall. When young emerge in spring, they resemble their parents, but are much smaller, lighter in color, and lack wings. In the South, two broods are produced each season.

MORMON CRICKET This serious pest of Western grains and other crops is partly controlled by insect parasites, small mammals, and birds. Gulls saved the crops of the early Mormon settlers from hordes of these wingless katydids that descended upon them. The large, clumsy insects devour everything in their path. Some Western Native Americans considered them a delicacy and ate them roasted. Small clusters of eggs are laid in the ground by the female.

AMERICAN COCKROACH 1.4″

AMERICAN COCKROACH Native to Southeast Asia and most common in the South, cockroaches live in houses, barns, and fields. They eat all kinds of food and sometimes destroy books, rugs, and clothing. They prefer moist, dark places and come out mostly at night. The egg case may protrude from the end of the female's abdomen for many days. This roach differs from the smaller German roach (pp. 152–153).

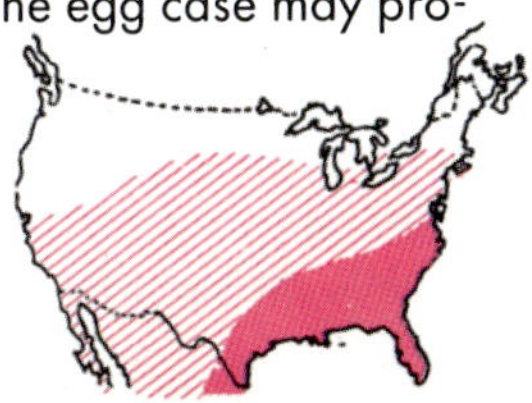

MANTISES These large, slender insects, generally called praying mantises or mantids, are apt predators, feeding mainly on other insects. If confined, mantises are likely to turn cannibal. They are colored a protective green or brown. Hard to see on foliage, they wait in ambush, snatching passing insects with their spiny forelegs. Mantids use their exceptional powers of vision and ability to rotate their heads to detect the movements of prey. In fall, after coupling, the female may eat the male. Curiously, the male's abdomen can continue to mate long after he has

lost his head. She lays several hundred eggs in a frothy mass that dries like hardened brown foam. Egg cases can be found in winter but will hatch prematurely if brought indoors. The young, similar to adults, are difficult to raise. The Carolina mantis is smaller than the others. It is one of 20 native species found most commonly in the South. The Chinese and European mantises, introduced here more than 50 years ago, are abundant throughout the East.

TERMITES Though sometimes called white ants, termites are not ants, and some are not white. Of some 2,000 species, only about 40 are found in this country. Many more are tropical. These highly socialized insects live in colonies composed of four distinct castes. The king and the queen, along with the winged termites that can become kings and queens of new colonies, form the first caste. The enlarged and almost helpless queen produces thousands of eggs over her lifetime, which may span a decade or two. Most of these hatch into whitish, blind

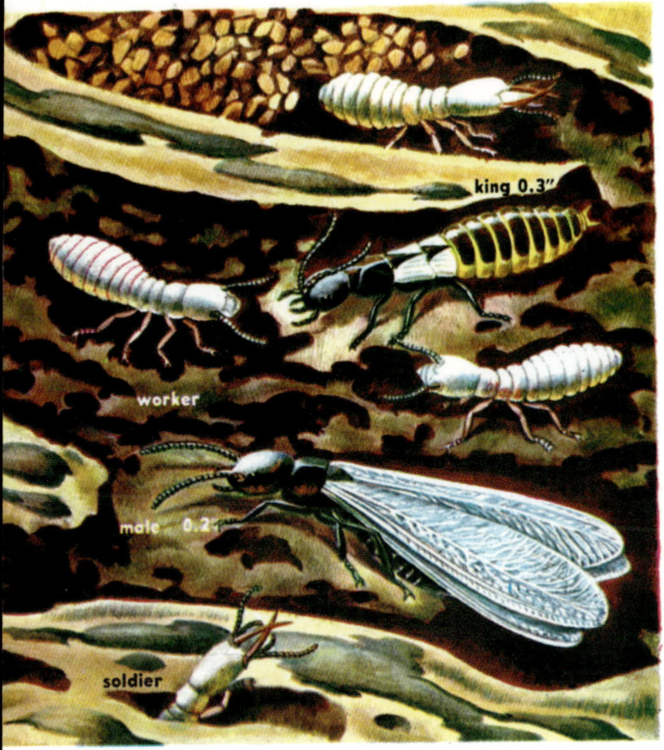

workers who make up the second caste. Soldiers with large heads and jaws, and nymphs, which take over the task of reproduction should the king or queen die, make up the last two castes. With the aid of protozoa living in their digestive tracts, termites feed on wood and do some billion dollars' worth of damage annually to buildings in this country. The young pass through six stages as they develop into adults. Some tropical termites build huge nests or mounds that are taller than a person.

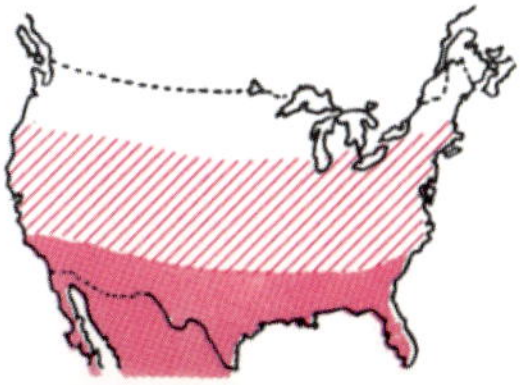

EARWIGS have short, leathery forewings and a pincer-like abdominal appendage. From abdominal glands earwigs may exude a liquid with a tar-like odor. They are nocturnal, spending the day in crevices or damp places. The legend of their creeping into ears of sleeping people is untrue. Although several species occur in North America, the most widespread is the introduced European Earwig. It feeds on dead organic matter, fungi, pollen, and plants in addition to insect prey.

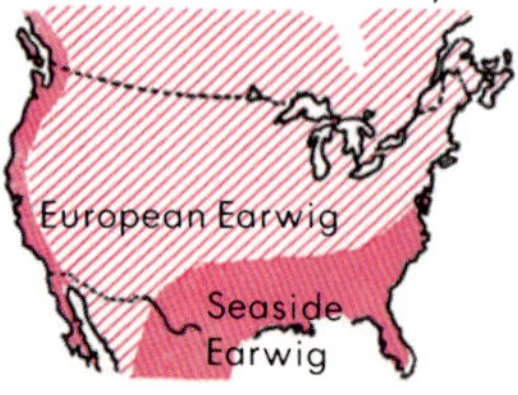

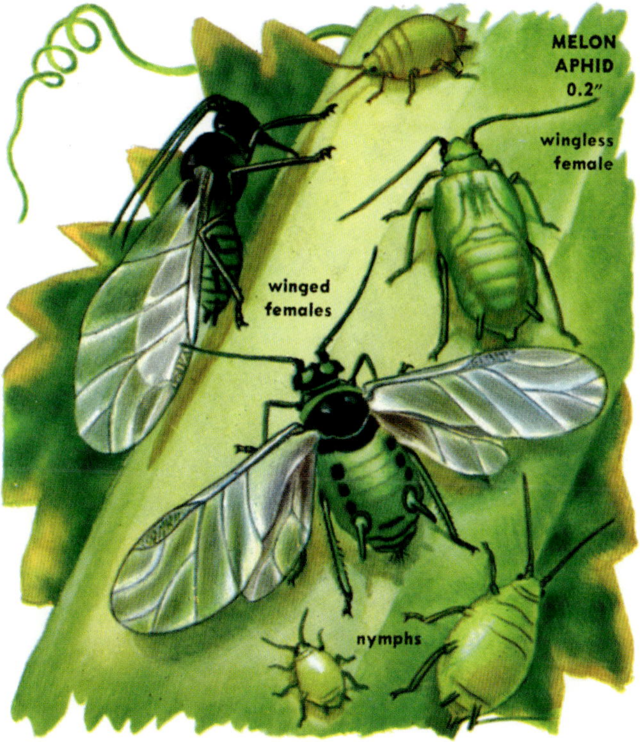

APHIDS are minute to small sucking insects, wingless or with transparent or colored wings. They are abundant on many plants, causing damage by sucking juices or transmitting viral diseases as they feed. They especially favor young growing shoots and flower stalks. Some form and live in galls. Most have complicated life histories. Only wingless females emerge from the eggs in spring. These produce generations of females all summer—sometimes a dozen. Winged females develop in fall. Their young are normal males and females, which, after mating, produce the eggs from which new aphids emerge in spring.

LEAFHOPPERS These attractive, slender, multicolored insects are often abundant on plants where they feed by sucking the sap. This causes wilting and injury to grape, apple, clover, beet, and other plants. Leafhoppers sometimes carry virus diseases from plant to plant and thus become serious pests. Leafhoppers exude "honeydew" as they feed. This somewhat sweet surplus sap attracts ants and bees, which feed on it. Leafhoppers are well known as prodigious jumpers. They are sometimes called dodgers because of the way they slip out of sight when ap-

proached. The female lays eggs in stems and leaves. Two or more generations are produced each year. Late eggs winter over and hatch in spring. Adults hibernate and emerge in spring also. The young pass through 4 or 5 nymph stages before they mature. Leafhopper populations in fields may reach as high as a million per acre. Of some 10,000 known species, about 2,800 are found in the United States.

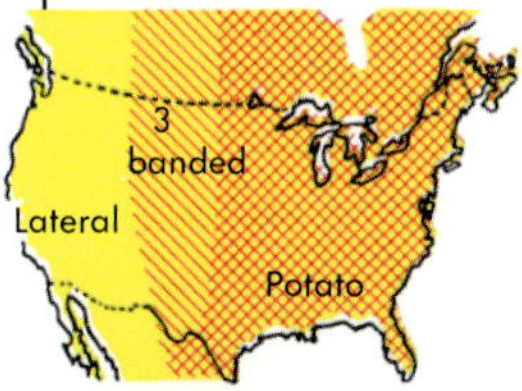

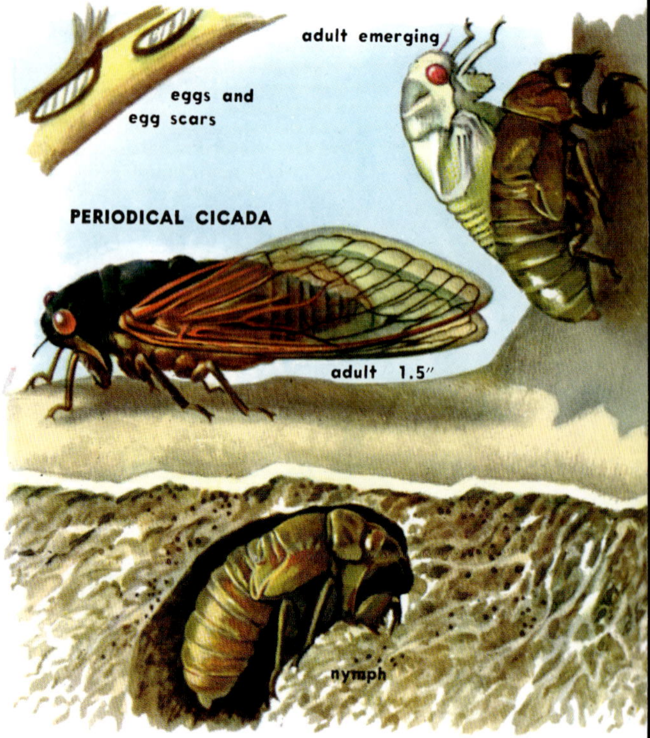

CICADAS, whose steady hum fills the summer air, are more often heard than seen. Males make the sharp sound with a membranous organ on the thorax. Ridges on the membranes make the sounds, which are amplified by the abdomen. Some species are called harvestflies because of their late summer appearance; the 13-year and 17-year cicadas are among our longest-lived insects. Most of our 75 species mature within a year. Some females cut slits in young twigs and deposit eggs in them. This sometimes causes great damage because the slit twigs break easily in

the wind. As the wingless young hatch, they drop to the ground, burrow in, and stay there 1 to 17 years (depending on the species and the latitude) as nymphs living on juices sucked from roots. The full-grown nymph climbs a tree trunk, usually during the night. Its skin splits down the back, and the adult emerges, leaving behind a distinctive shell. In most species, adults live a few weeks—long enough to mate and lay eggs. "Broods" of periodical (13- and 17-year) cicadas emerge en masse.

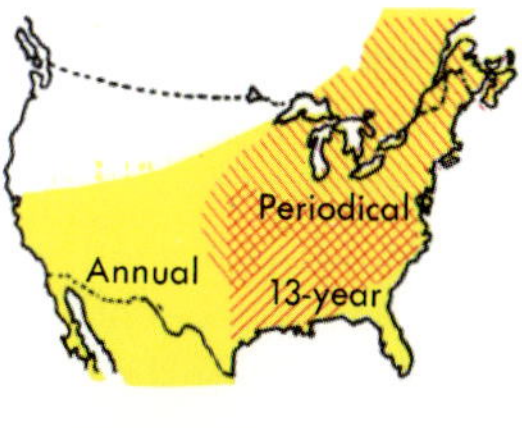

TREEHOPPERS The common green and brown treehoppers are small, winged, sucking insects with peculiar shapes. They live on many plants, feeding on sap. Because of their protective color and form, they are usually noticed only when moving. Nearly 200 species are known in this country. Eggs are laid in stems and buds, sometimes causing minor damage. Eggs hatch the following spring. Many species of treehoppers are protected from their natural enemies by ants, who benefit by taking a honeylike solution from the nymphs to eat.

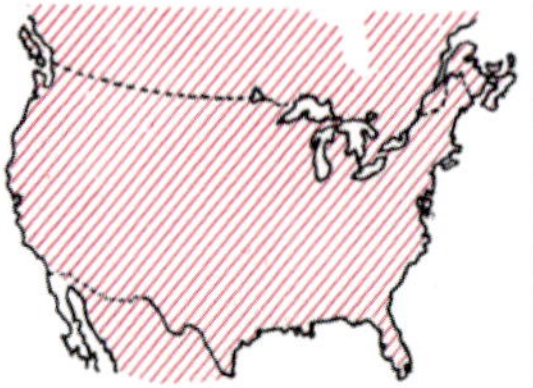

SPITTLEBUG Spittlebugs use their rear legs to whip up an abdominal secretion into a gluey froth. Females use it on stems and grasses to cover their eggs, nymphs to cover themselves while feeding. Open the small mass of bubbles and you are likely to see the small, dull, squat insect inside. Spittlebugs are also called froghoppers because the adults are great jumpers, hopping from plant to plant. Some kinds occasionally injure pine trees and garden plants.

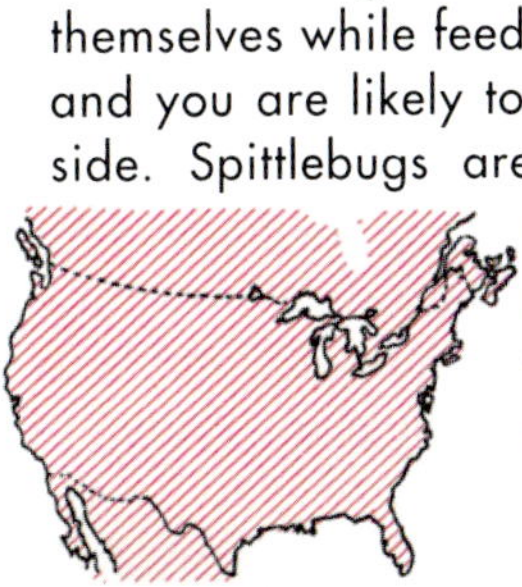

SAN JOSE SCALE 0.1″

SCALE INSECTS are a large group of small sucking insects. Individually minute to small, these insects live in colonies which often cover branches, twigs, and leaves of the plants on which they feed by sucking juices. Species differ markedly in appearance. Many have a scale-like covering and are immobile when mature. Other species lack scales, but have poorly developed legs and move very little. Some are covered with a "honeydew" secretion eaten by bees and ants. Males are smaller and differ from the females; when mature they have small wings. Scale in-

sects attack and injure citrus, apple, other fruit trees, and greenhouse and ornamental plants. They can be difficult to control because they are covered by a waxy plate. Ladybugs, certain small wasps, and other natural enemies have been used successfully in fighting them. Reproduction is complicated. Most scale insects spend winter as eggs, which the female deposits under her shell before she dies. The eggs hatch in spring and the young move to fresh growth before they settle down and form their scale. Some species produce several generations of females before normal sexual reproduction takes place.

STINK BUGS AND SHIELD BUGS There are several hundred species of stink bugs and shield bugs in this country. All have a flattened, shield-shaped body. Most suck plant juices but some are predators of other insects. Most are colored green or brown, to match their environment, and are not easily noticed. A black species common on blackberries and raspberries is so well concealed it is sometimes eaten. The colorful harlequin cabbage bug is an exception, even to its unusual eggs. The odor, which comes from two glands on the thorax, gives stink bugs

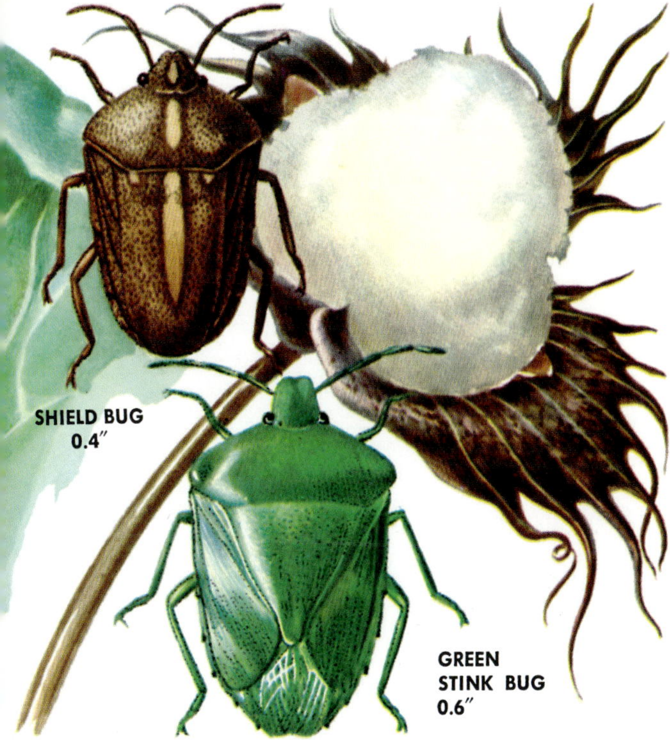

their name, but it is also characteristic of a number of other bugs. Birds are not bothered by the odor and commonly feed on these insects. The harlequin cabbage bug and several other species are occasionally destructive to garden crops. The shield bugs are very similar to stink bugs. In these species, the shield, which develops from the thorax, is so large that it covers a good part of the abdomen.

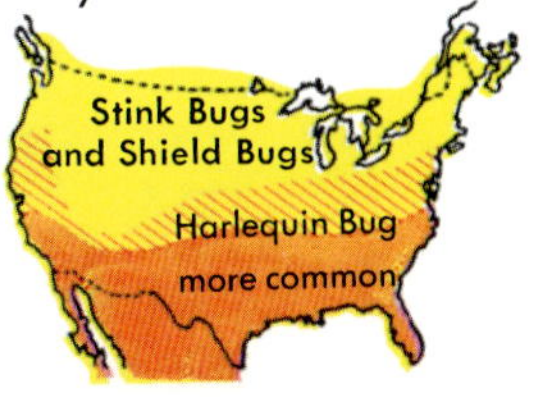

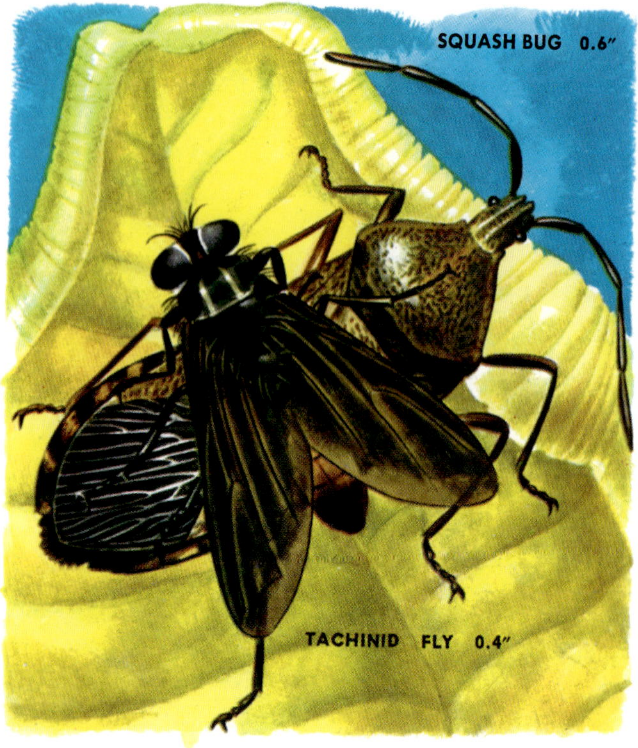

SQUASH BUGS cause considerable damage to squash, pumpkins, gourds, and related crops by sucking juices from leaves and stems of young plants. The bugs have a strong, offensive odor. Eggs, laid in late spring, hatch in about 2 weeks. The attractive nymph is green, soon turning brown or gray. Adults hibernate over winter. A tachinid fly, which lays its eggs in nymphs and adult squash bugs, parasitizes these pests and helps reduce their numbers.

MILKWEED BUGS These black and red or orange bugs are related to the tiny destructive chinch bug (p. 49). About 200 species are grouped in the same family with the milkweed bugs, but most are much smaller and less attractive in color and pattern. Milkweed bugs feed on all varieties of milkweed and are of no economic importance. Adults that hibernate over winter produce young in late spring. The nymphs mature and breed by late summer.

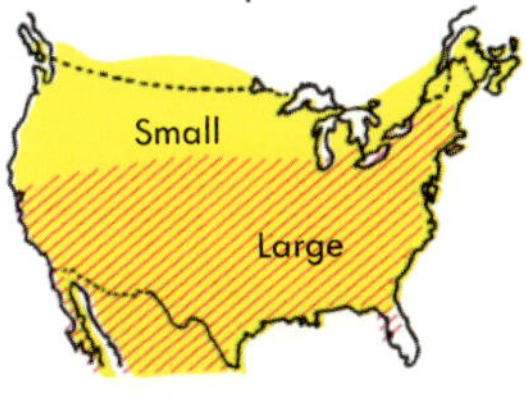

AMBUSH BUG These small, oddly shaped predators form a minor group of some 25 species. They lie concealed in flowers and grab any small insects which come by. Their front legs are modified for holding their victims; their mouth for piercing and sucking. Ambush bugs have little economic importance, but their bizarre forms, unusual feeding habits, and uncanny ability to blend with flowers make them interesting to study.

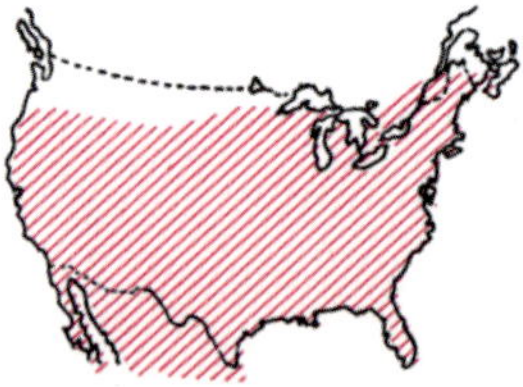

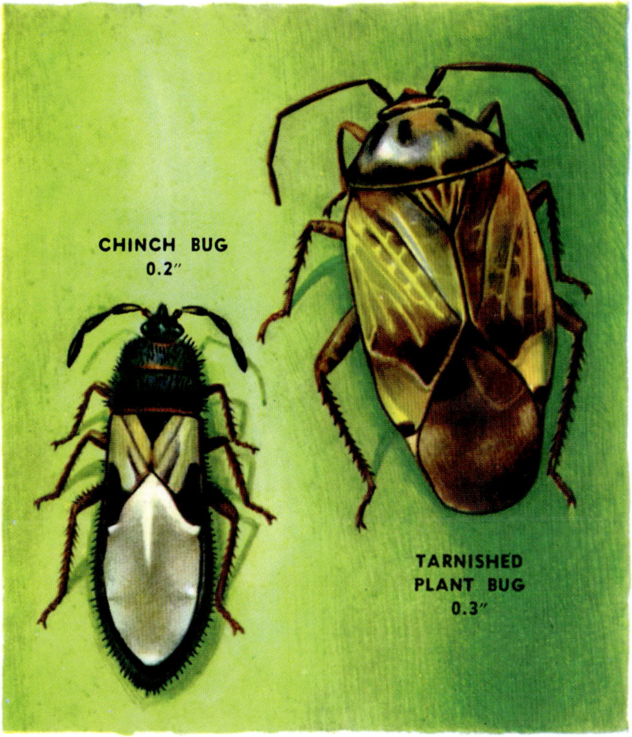

CHINCH BUG Though small, almost minute, chinch bugs reproduce so rapidly they overrun grain fields, destroying the crops as they feed on plant juices. The annual damage in this country runs into millions. About 500 eggs are deposited in grass or grain. Nymphs are red, becoming gray or brown with age. Two or three generations may develop in one season. The tarnished plant bug, somewhat larger, and of a related family, is destructive to many kinds of fruits.

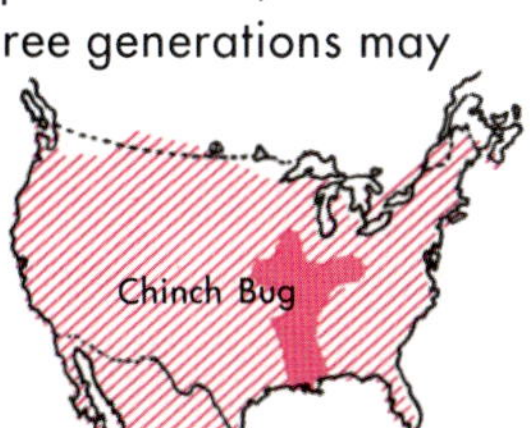

AQUATIC BUGS are found in nearly every pond and stream. A few species are marine. All are remarkable for their adaptations to life on or below the surface. Nearly all are predaceous, attacking other insects, snails, small fish, etc. In turn, these insects are food for larger fish and water birds. The water striders, taking advantage of surface tension on water, stay on the surface without breaking through. They skate along with remarkable speed. Other water bugs spend most of their time under water. Some carry down air on the surface of their bodies and

use this for breathing. Other species use air that is dissolved in the water. In most cases, the young resemble the adults and mature after a series of nymph stages. The water-boatmen, common in pools, swim erratically. The backswimmers, as their name indicates, swim on their backs, but can also fly. The giant water bugs, sometimes 2 inches or more in length, prefer quiet water. Since their bite can produce a painful swelling, they are best not handled.

SHORT-NOSED CATTLE LOUSE 0.1+"

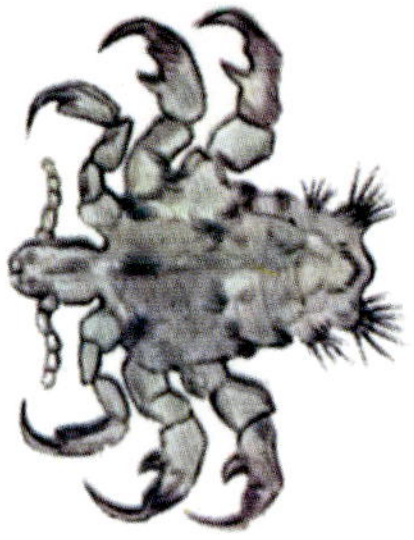

CRAB LOUSE 0.1—"

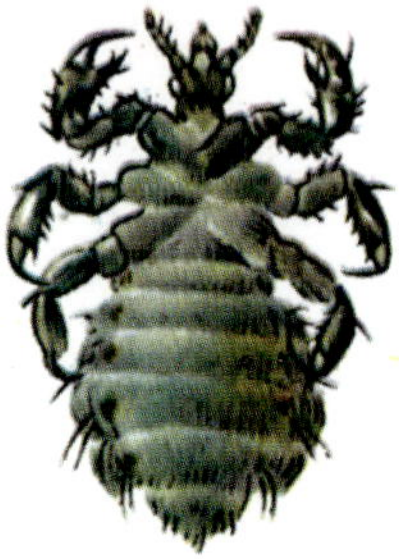

BODY LOUSE 0.1+"

LICE are minute, wingless insects that live and breed on their hosts. All are parasites; some carry disease. Chewing lice (bird lice), a distinct group, feed on hair, feathers, and fragments of skin. The sucking lice consume the host's blood directly, by means of sucking mouth-parts. The hog louse (¼ in.) is the largest of this group. The head louse infects humans and is known to transmit typhus, trench fever, and relapsing fever. Six to 12 generations of lice may mature annually. Young, similar to adults, develop rapidly.

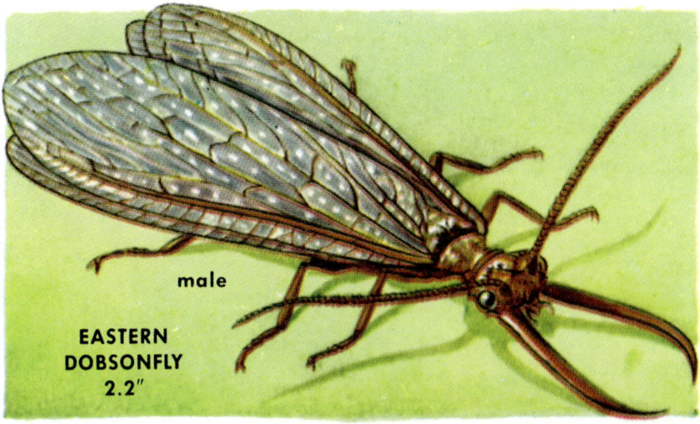

DOBSONFLY The ferocious-looking adult male is harmless. The long mandibles are used in mating, which ends its short adult life. The female lacks these exaggerated mouthparts; and therefore is suitably equipped to give a serious nip. She lays a mass of thousands of eggs on plants overhanging a pond or stream. The larvae emerge, drop into the water, and spend the next three years or more feeding on smaller water life. Fishermen prize the large larvae, called hellgrammites, as bait.

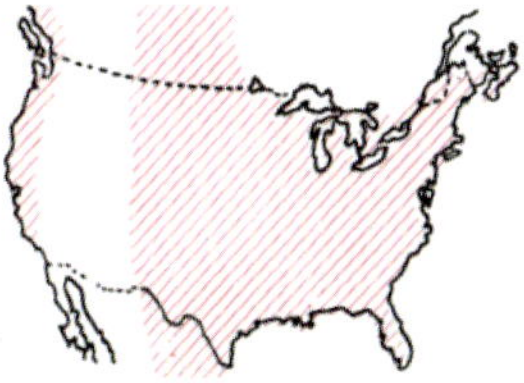

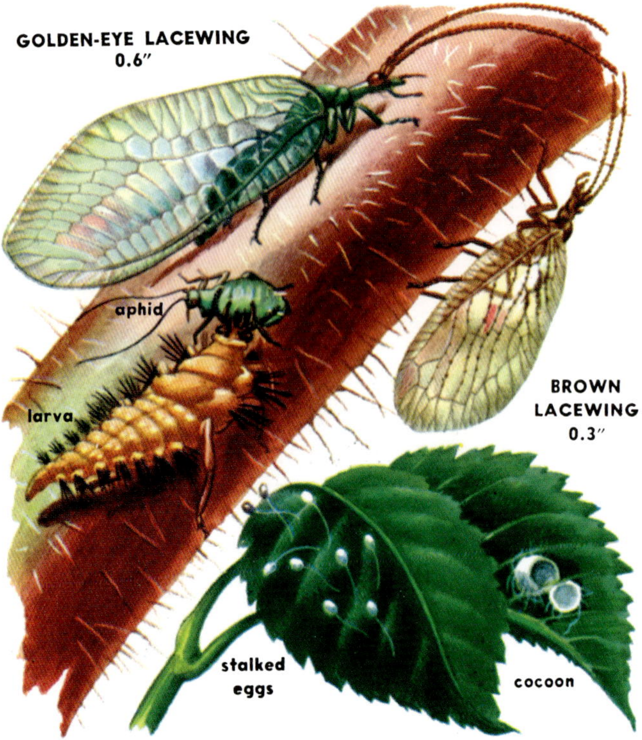

LACEWINGS The larvae of North America's 50 or more kinds of green lacewings are voracious predators. They feed on aphids and other destructive insects, earning the name aphid lion. Eggs have stalks that give them some protection from older larvae. Larvae spin silky cocoons from which they emerge as delicate thin-winged adults. Courting adult lacewings "sing" by vibrating their abdomens, sending sound waves through the surfaces on which they are standing.

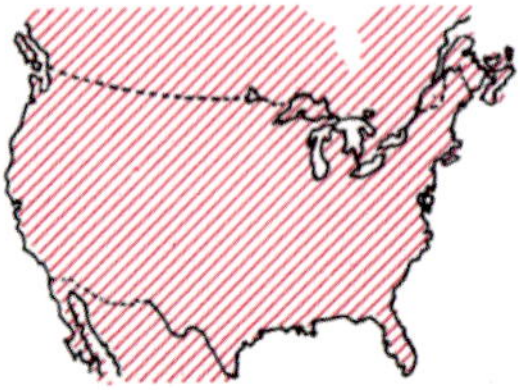

ANTLIONS are so named because the larvae of members of this family have odd feeding habits. Eggs are laid in sand. When one hatches, the larva digs a pit in sand or sandy soil and lives almost completely buried at the bottom. Should an ant or other small insect tumble in, it is seized in powerful jaws, poisoned, and sucked dry. As the larva, or "doodlebug," matures, it builds a silken cocoon, in which it pupates. The adult resembles a damselfly, but with short, knobbed antennae; they frequently come to lights.

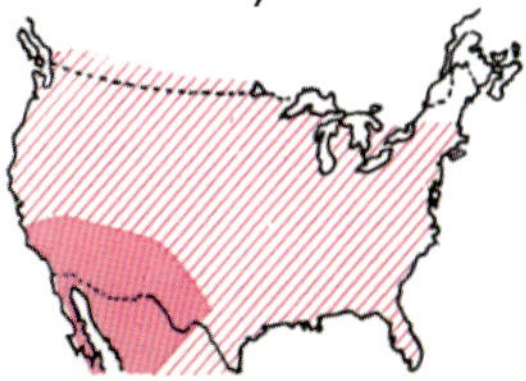

TIGER BEETLES These handsome beetles are often seen on sunny days. They are widely distributed and quite common, but agile, wary, and difficult to catch. Eggs are laid in the soil. The flat-headed, hump-backed, predatory larvae dig deep burrows and wait at the openings to catch passing insects in their powerful jaws. Markings on some tiger beetles living on beaches or sandy areas blend with their surroundings. Some species are rare or locally extinct; several are protected by state and federal laws.

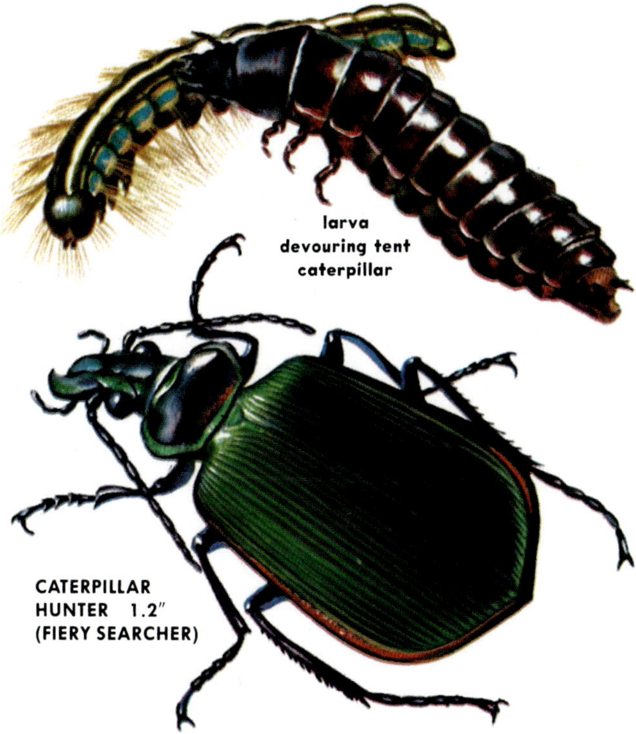

GROUND BEETLES include some fairly large and very attractive species, such as the caterpillar hunter above. This family is large (some 2,000 American species) and includes the tiger beetles (p. 56). Most are predators, feeding on insects and other small animals. Some adults squirt a sour-smelling fluid at predators or on unwary collectors. The fluid sprayed by a bombardier beetle attains the temperature of boiling water before it is literally blown out of the back end of the abdomen.

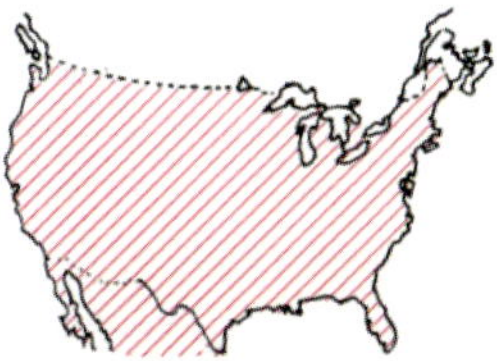

CARRION BEETLE 0.7″ **BLACK CARRION BEETLE 0.5″**

CARRION BEETLES form a family of over 100 species, in two groups: the carrion and the burying beetles. Carrion beetles are smaller, flattened insects. Both larvae and adults have similar feeding habits. Some are scavengers, feeding on decaying animal matter; some are predators, feeding on worms and insects; and a few eat plants. The

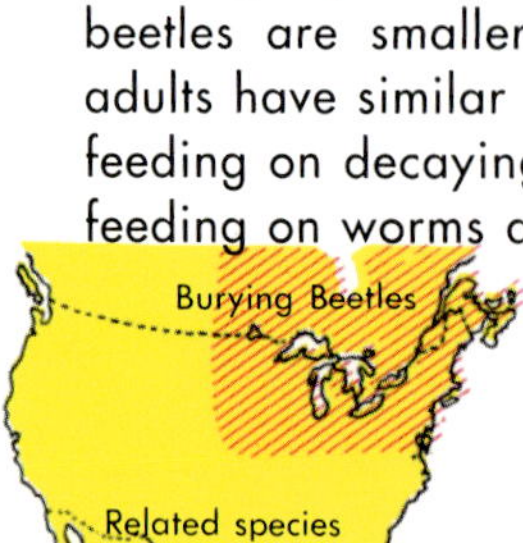

larger burying beetles, which are sometimes brightly colored, dig under the carcass of a small animal until it falls into the hole and is actually

AMERICAN BURYING BEETLES 1.2″

HAIRY BURYING BEETLE 0.7″

HAIRY ROVE BEETLE 0.8″

buried. Eggs are deposited on the corpse, which serves as food for the larvae. The larvae of some species develop rapidly, reaching maturity in about a week. The American burying beetle is the largest North American burying beetle and an endangered species. Rove beetles, from another family, have short wing-covers and are also scavengers or predators. Over 3,000 species are reported for this country. Some live in fungi or in ant nests. A few squirt a malodorous mist when disturbed.

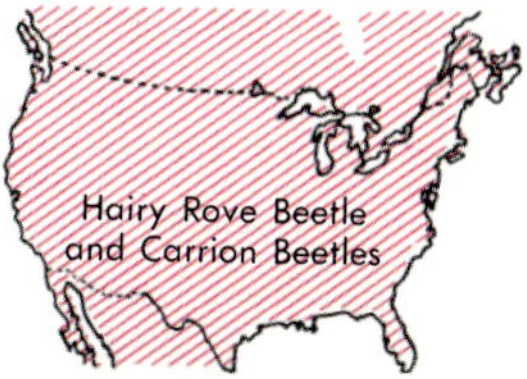

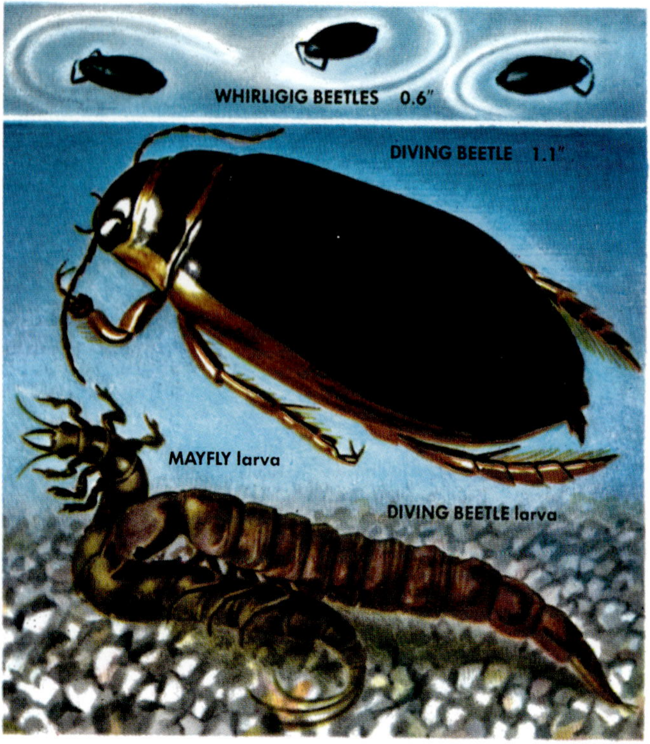

WATER BEETLES are not to be confused with water bugs (pp. 50–51), though they are often found in the same habitat. Whirligig beetles, true to their name, whirl or swim at the surface. The eye is divided into an upper and lower portion, thus the larvae can see above and below the surface simultaneously. They dive when disturbed and are also good fliers. Eggs are laid on water plants; larvae feed on water animals. Eggs of diving beetles are deposited in the tissues of water plants. They hatch into larvae commonly known as water tigers, prey upon water insects, small fish, tadpoles, even one another. After a month or

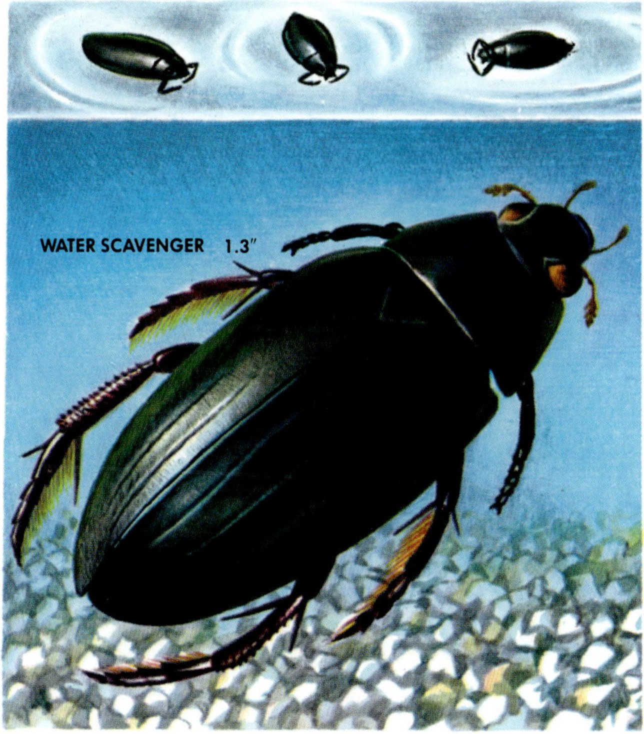

so, larvae leave the water and pupate in the ground. When they dive underwater, adults carry a supply of air at the tip of their abdomen.

Water scavengers (225 species) include our largest water beetles, some over 3 inches long. Eggs are laid in a silken wrapping, attached to a floating leaf. Predaceous larvae feed upon a myriad of pond animals. Pupae form in soil by late summer, and emerge in about 2 weeks. They carry air as a film on the underside of the body.

FIREFLIES AND GLOWWORMS are of little economic importance, but add to the pleasure of a summer night. Thousands of these very unusual insects flashing in unison are a breathtaking sight. They are not flies at all, but soft-bodied beetles belonging to two families. About 130 species are known in this country, and many more, some even more marvelous, are found in the tropics. The light-giving property, or luminescence, is not confined to adults. In one family, the glowworms, the eggs and larvae also glow. Females of some species are short-winged or wingless. The pattern of flashing allows individuals to find partners of the correct species and sex for mating. The larvae live underground or in rotted wood or rubbish, feeding on

worms and snails. Adults are also predators, but some common fireflies may not feed at all in the adult stage.

In the species shown, the last segment of the abdomen contains the light-producing tissue. It is very fatty and contains a network of nerves and airtubes. These tubes carry the oxygen necessary to produce light. In most reactions, more heat is produced than light, but here, heat production is negligible. Put some in a jar, turn out the lights, and watch the action. Be sure to release your quarry afterwards. They are difficult to maintain in captivity.

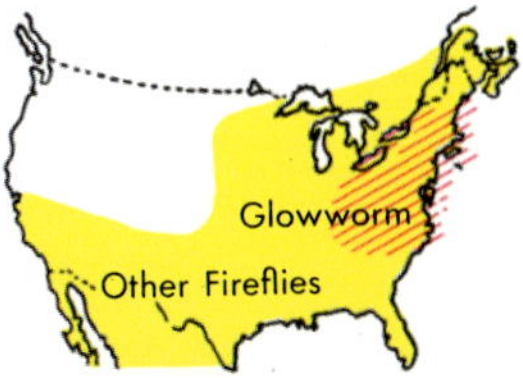

LADYBIRD BEETLES are probably the best known and most valued of our beetles. There are some 350 species in this country. Both larvae and adults of many species feed on aphids. In California, where aphids and scale insects cause serious damage to citrus trees, native and imported ladybird beetles have been used successfully to control these pests. When the Cottony-cushion Scale from Australia (pp. 42–43) spread through California orange groves, the entire industry was threatened. An Australian Ladybug, which feeds only on the scale, was imported. Within a few years, the scale was under control. When mature, ladybird beetles pupate in their last larval skin. Contrary to childhood myth, the number of spots on a lady-

beetle has nothing to do with its age. Some species are entirely black, or black with orange or yellow spots, and others, such as the recently introduced Halloween Ladybug, vary from all black to spotted to all orange. Adults may assemble by the thousands before cold weather sets in and hibernate together. Prior to the onset of winter, Halloween Ladybugs may enter homes and other buildings by the hundreds. These can be saved by vacuuming the insects into an unused bag and then hanging the bag in an outdoor shed or woodpile. Many once common, native lady beetles are now rare, perhaps because of competition with the many introduced foreign species.

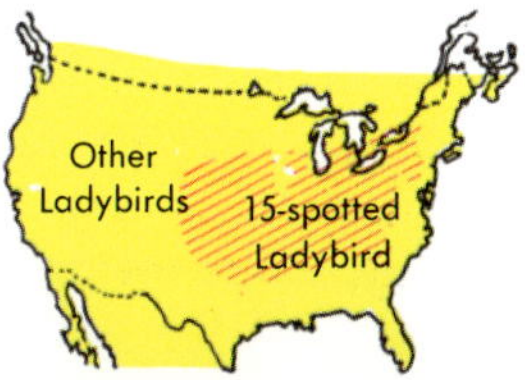

SCARAB BEETLES form a large family totaling more than 30,000 species. Well over 1,300 are found in this country, including the rose chafer, Japanese beetle, and May beetle. Many are scavengers, adapted for living in or on the ground. Larvae are the large white C-shaped grubs found in rotted wood or rich soil. Of the many scarabs, the dung beetles and tumblebugs are among the most notable. Adults laboriously roll a ball of dung to a site, lay their eggs inside the ball, and bury it. Their young feed on the

EASTERN HERCULES BEETLE female

OX BEETLE 1.0″

EASTERN HERCULES BEETLE male 2-2.5″

dung after hatching. The ferocious-looking rhinoceros beetles and their relatives, the ox beetles, are the largest North American scarabs, attaining lengths of more than 2 inches. Much larger species can be found in the tropics. All are harmless. Males have more prominent horns than females. Members of this family were held sacred by the ancient Egyptians and are prized by many collectors.

ROSE CHAFER This slim, hairy beetle is one of the many scarab beetles, an enormous family of over 30,000 species, diverse in size and appearance. Some are pests, while other groups, like the dung beetles, are important recyclers. The rose chafer feeds on roses, grapes, and other plants. Adults appear in late spring and early summer, eating both leaves and flowers. Larvae burrow into the ground, where they feed on roots, especially those of grasses.

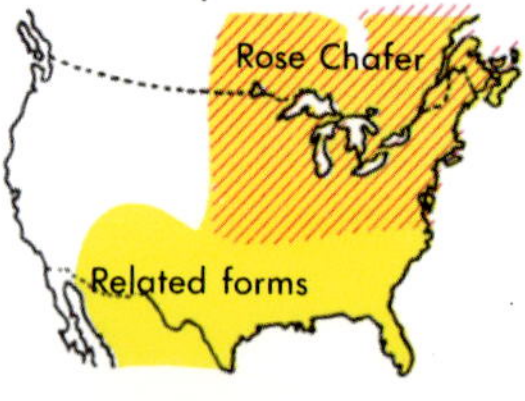

JAPANESE BEETLE 0.4″

JAPANESE BEETLE When these Japanese insects were first discovered on plants in New Jersey in 1916, experts could scarcely find a dozen. Now thousands can be collected daily. The small, white grubs feed on the roots of grasses, damaging lawns. Larvae dig deep for winter and pupate the following spring. Adults emerge in midsummer and feed on the leaves and flowers of many plants, especially roses and other flowers. After mating, eggs are deposited in soil.

MAY BEETLE 0.9″ and larva

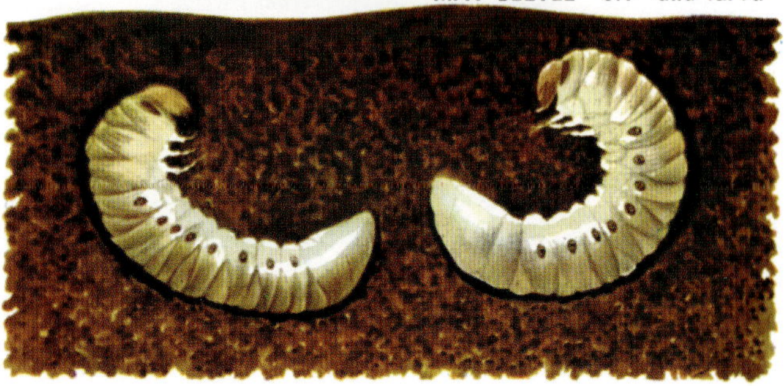

MAY BEETLES or June bugs form a group of over 100 widely distributed American scarab beetles. White eggs are laid in an earth-covered ball amid roots. When they hatch, the white grubs feed on the roots for 2 or 3 years. They pupate underground in fall, and adults appear the following spring. Adults feed on leaves of many common trees. They are attracted to electric lights. Birds and small mammals, such as moles, skunks, and even pigs, root out the grubs and eat them.

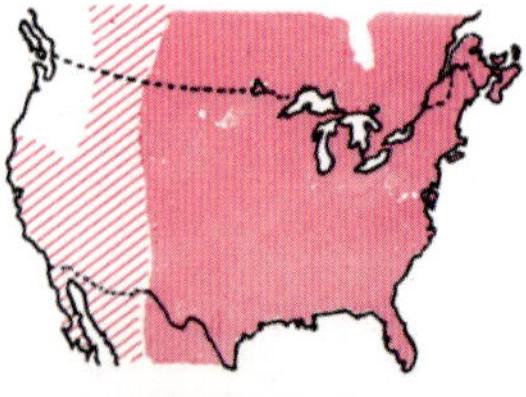

GREEN JUNE BEETLE 0.9" and larva

GREEN JUNE BEETLE Often called the figeater, this scarab beetle feeds on many plants, eating roots, stems, and leaves. Larvae are found in soil or manure. They move by pushing against the backward directed bristles instead of using their short legs. Adults may occur in large numbers, making a loud buzzing which is somewhat similar to the sound made by a bumblebee. These insects are more common in the South, where the adults damage apricots, figs, grapes, melons, and other fleshy fruits.

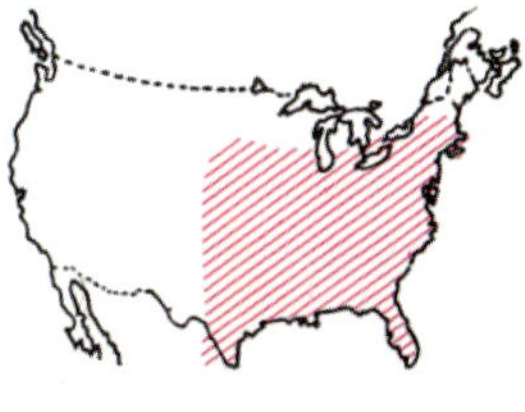

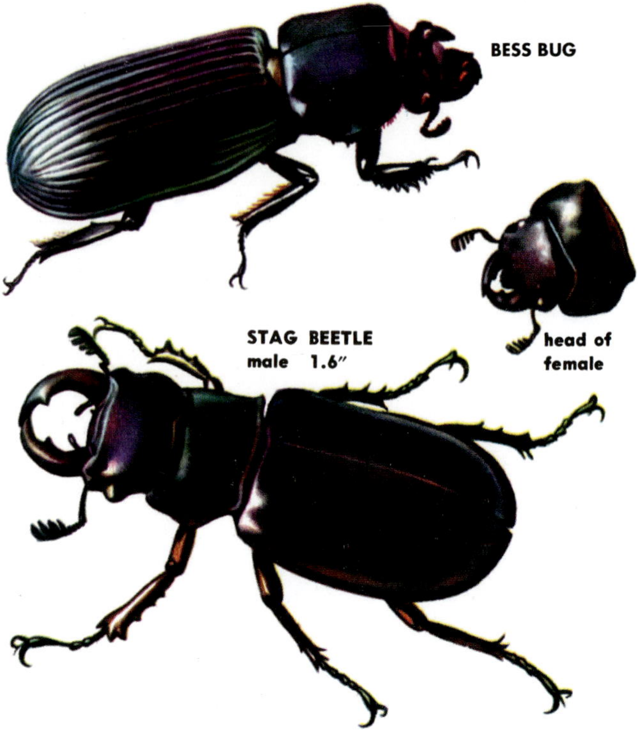

STAG BEETLES AND BESS BUGS are medium to large beetles closely related to scarabs. All are harmless. Stag beetles get their name because the male's huge mandibles resemble the antlers of a stag. The female's mouth-parts are much smaller. Large white stag beetle grubs are found in rotted wood—more commonly in the South than elsewhere. Large colonies of bess bug larvae and adults are often found in burrows in rotted logs. They make noise by rubbing their wing-covers or legs.

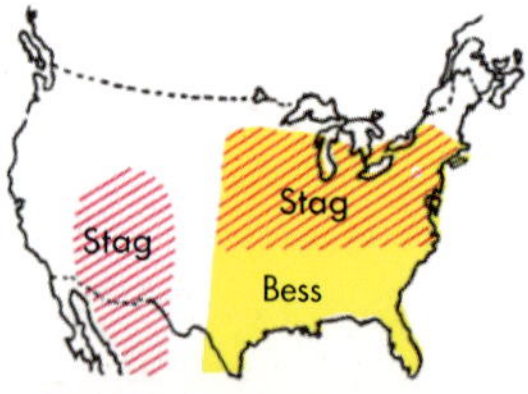

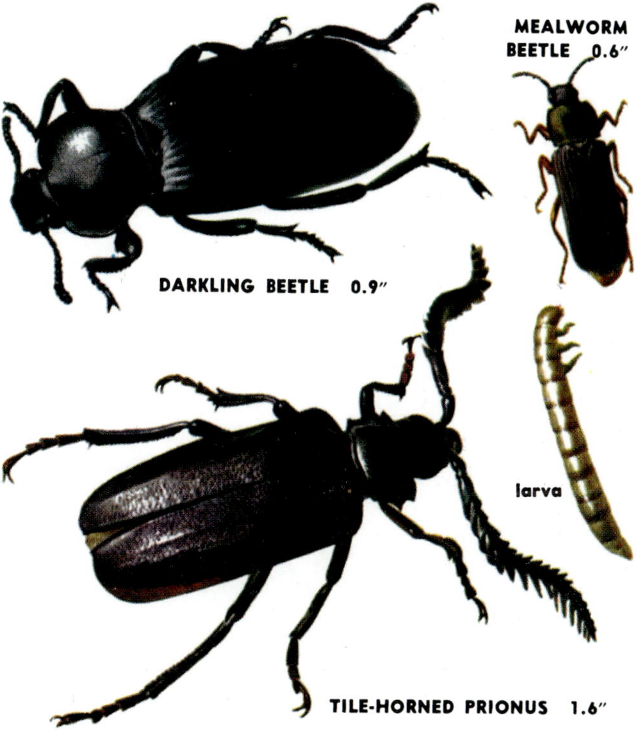

DARKLING AND PRIONUS BEETLES There are about 700 darkling species found in the U.S., many in arid areas. The larvae of a few species are the "mealworms" used for feeding pets or fishing. Some species feed on stored grain. Prionus beetle larvae feed on the roots of fruit and ornamental trees and other plants. Adults can be identified by the overlapping plates that form their antennae. Prionus beetles are actually longhorn beetles. (see p. 75)

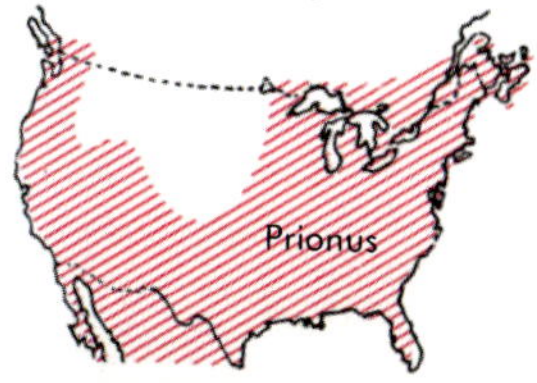

CLICK BEETLES form a family of some 5,000 American species. The eyed click beetle is a large and striking example. If a click beetle falls and lands on its back, it lies quietly for a period. Then, with a loud click, it flips into the air. If it is lucky, it lands on its feet and runs away; otherwise it tries again. The larvae of click beetles, known as wireworms, live in the ground or in rotten wood. Many click beetle larvae feed on roots, sometimes injuring potatoes and other crops. Some eat other insects.

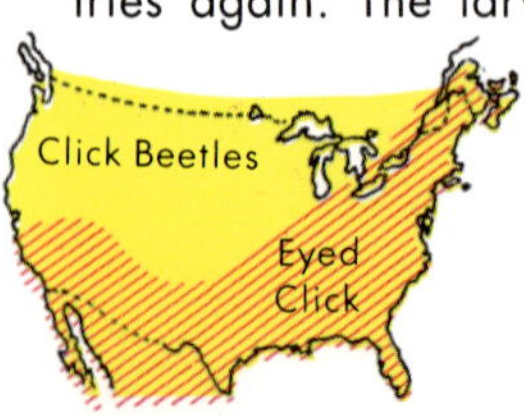

LONGHORNED AND FLATHEADED BORERS Larvae of many beetles bore into trees, but longhorned borers (represented by the locust and elder borers and the pine sawyer) and flatheaded borers are occasional pests. Larvae of longhorned borers (over 1,000 species) usually channel deep into the wood. Flatheaded borers (at least 500 species in the U.S.) feed mostly just beneath the bark. Many borers are eagerly sought by collectors. Some stridulate (squeak) when held or disturbed.

BEETLES COMMON IN GARDENS

Some beetles are significant pests in gardens as well as on farms. Because of selective breeding, many garden plants and crops have weakened natural defenses against insects such as beetles. This makes these plants ideal targets for attack. Growing lots of the same crop plant in a single area, as we do in modern farming, also enhances the possibility for severe infestations from pest species. Wild relatives of garden plants and crops are often more resistant to insect attack.

MEXICAN BEAN BEETLE is closely related to the common ladybird, and is one of two North American species within the ladybird family that feed on plants. Eggs are laid on the undersides of leaves. Spiny, yellow larvae eat the soft leaf tissue, leaving the veins behind. They eat pods too, stripping a plant in short oder. Adults have similar feeding habits. Bean beetles feed on wild and cultivated members of the pea family—peas, beans, alfalfa, and soybeans.

COLORADO POTATO BEETLE is an example of how a relatively unimportant insect can change its role as the environment changes. This beetle was once native to the Rockies, living on nightshade and other wild members of the potato family. When settlers began to grow potatoes, the beetle prospered and spread throughout much of the U.S. Eggs are laid in clusters on the leaves, which both larvae and adults eat.

STRIPED BLISTER BEETLE or striped potato beetle has interesting relatives that parasitize bees. This species has a complex life history with unusual larvae. As an adult it feeds on potatoes, tomatoes, and related plants. Other species feed on goldenrod, alfalfa, clover, and other wild plants.

MEXICAN BEAN BEETLE
0.3″
eggs and larva

COLORADO POTATO BEETLE

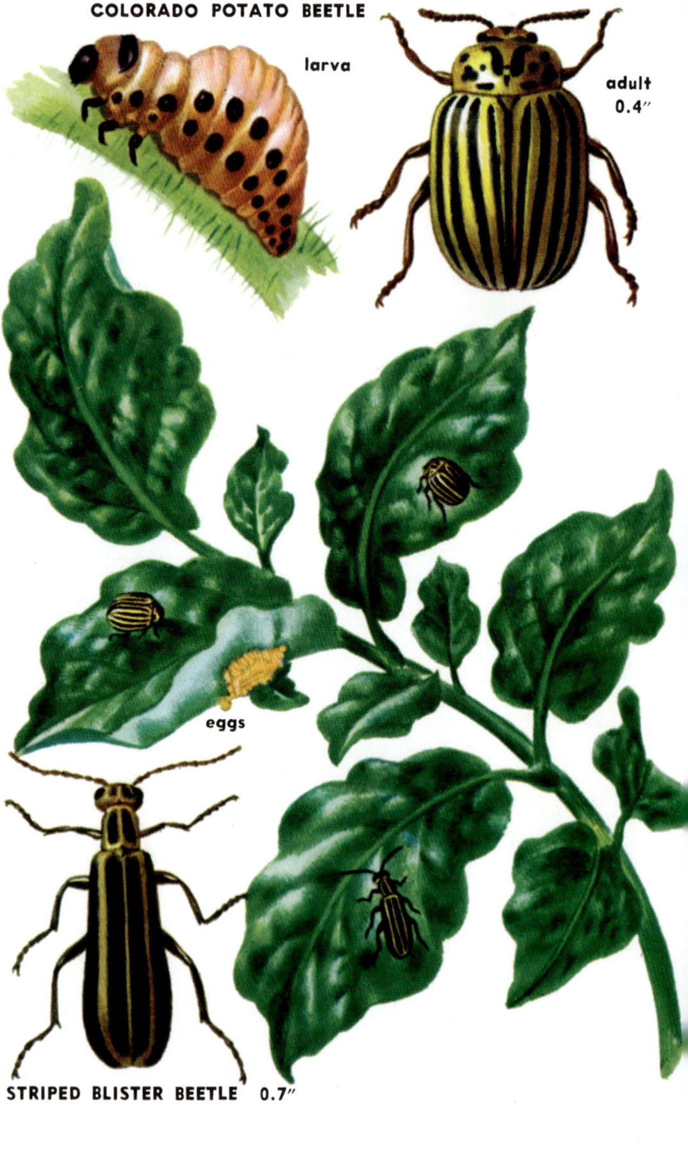

STRIPED BLISTER BEETLE 0.7″

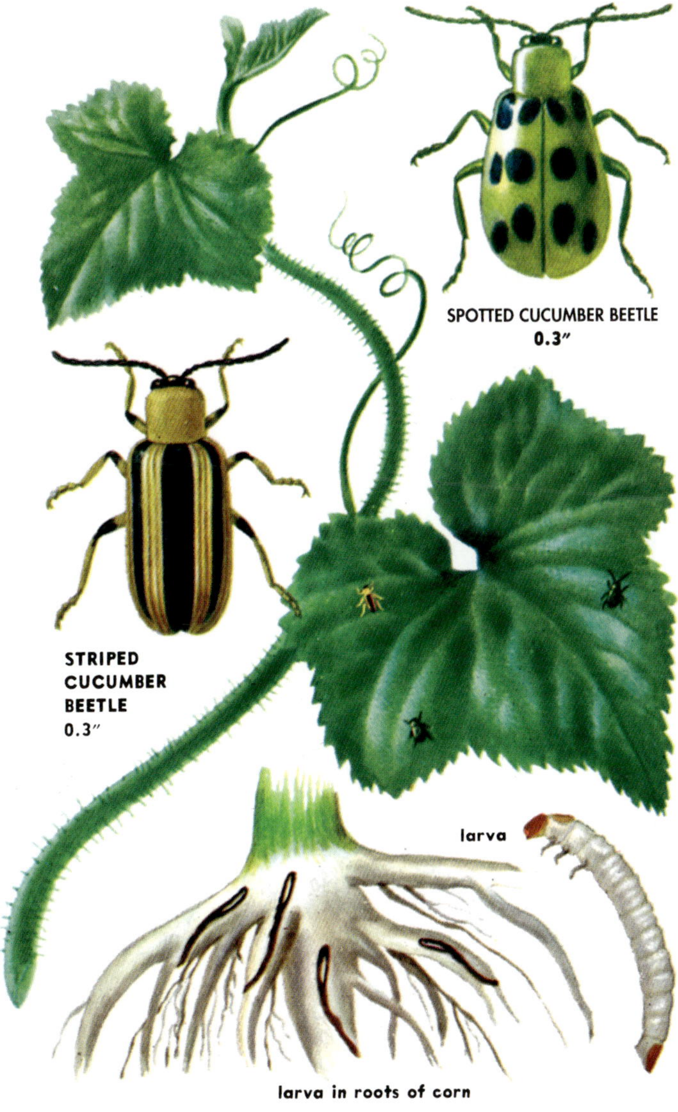
SPOTTED CUCUMBER BEETLE
0.3″
STRIPED
CUCUMBER
BEETLE
0.3″
larva
larva in roots of corn

CUCUMBER BEETLES The striped and spotted cucumber beetles are common garden pests. The larvae of the former attack roots, and the adults eat leaves of cucumber, squash, and related plants. The spotted cucumber beetle feeds on many other plants besides those of the squash family. It appears early in the season and stays late. In the South, the larvae bore into the roots of corn, oats, and other grasses.

ASPARAGUS BEETLES Two leaf beetle species are known to eat asparagus. Adults hibernate in the ground, emerging in spring to feed on young shoots. Eggs are soon laid and larvae attack shoots, "leaves," and fruit, stripping the plant. The cycle takes only about a month; there are several broods a year. Both species, the common asparagus beetle and the spotted asparagus beetle, were introduced from Europe, one about 90, and the other about 70, years ago.

CONTROLLING GARDEN PESTS Most insects found on plants in your garden are not pests. Many are beneficial. Therefore, it is important to identify the insect causing the problem and to learn something about its eating habits. If you cannot, seek aid of a nursery or your county agent, college of agriculture, or the U.S. Department of Agriculture. Most have Web sites with helpful information (see p. 9). Once you know which insect is being a pest, proper control measures can be obtained from the same sources. State and Federal agricultural agencies often have pamphlets and Web pages devoted to controlling garden pests. It is best to get the most up-to-date information that you can because pest control methods are being constantly developed to meet new conditions.

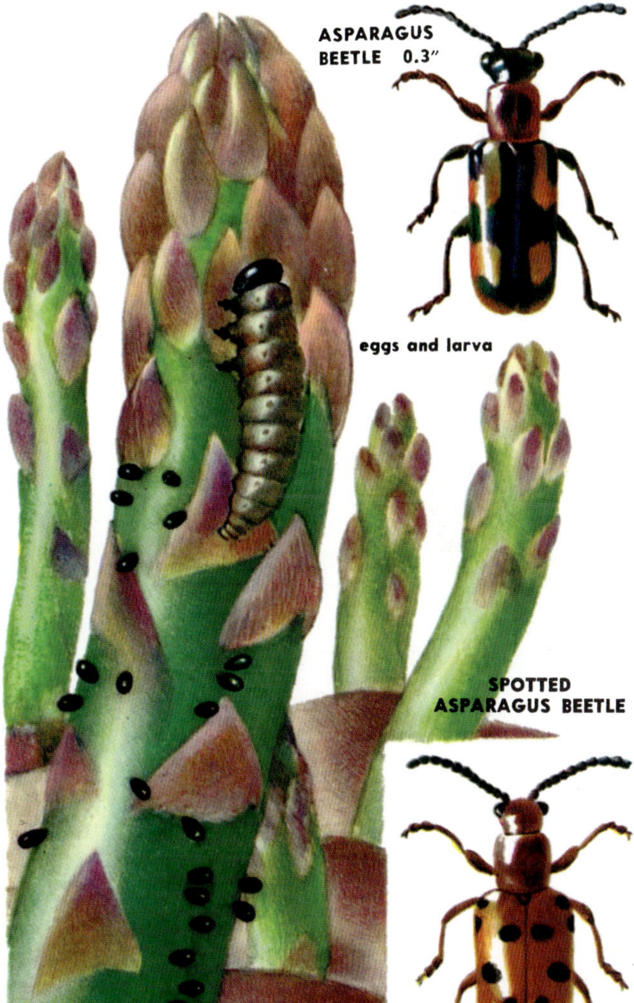
ASPARAGUS
BEETLE 0.3″
eggs and larva
SPOTTED
ASPARAGUS BEETLE

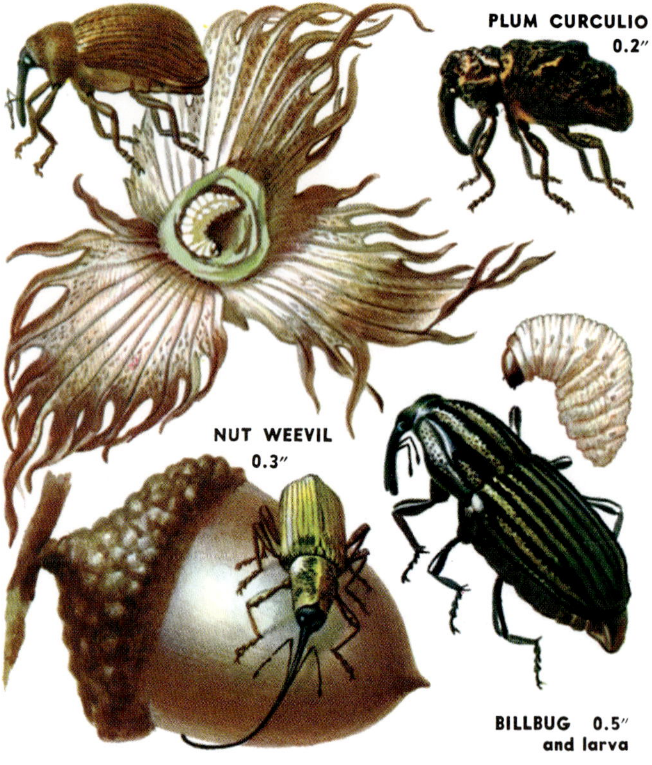

WEEVILS are small beetles with a downward-curving beak or snout. They represent the largest family of insects in the world, with more than 80,000 species. All U.S. weevil species are plant eaters; most live inside a plant, but a few are found on leaves. Grain weevils, important pests of grain, have a life cycle of only 4 weeks. Some feed on the roots, others on stored grain. Boll weevils are a danger to cotton crops. The plum curculio damages peach, cherry, and plum trees. Nut weevils are found in acorns and all other edible nuts.

CADDISFLIES include over 1,300 North American species. Larvae live in fresh water, most in portable tube-shaped cases that they make by fastening sand grains, plant material, or bits of aquatic debris together with silk. Other larvae build stationary nets out of silk that capture food particles drifting downstream. Some larvae are free-swimming. Larvae turn into pupae in water, then crawl out of the water and molt into adults. Adults are usually drab and moth-like, with hairy wings. Larvae are important ecological indicators and are used to determine water quality. All stages of caddisflies are important food for fish and other organisms.

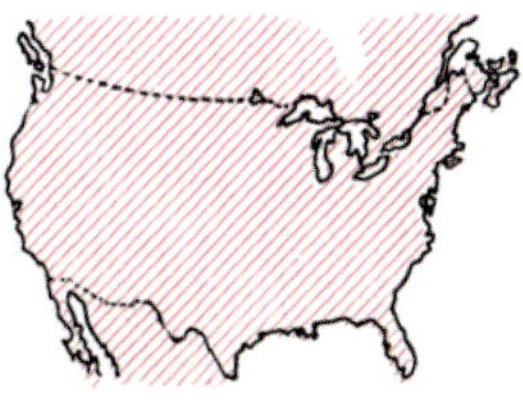

antennae of butterfly (left) and moth

BUTTERFLIES AND MOTHS The largest, most attractive, and best-known insects are grouped together in the order Lepidoptera, the butterflies and moths. In North America, the order includes 5 families of butterflies and 75 of moths. About 14,000 species are known to occur north of Mexico. All, except very few, have two pairs of wings. These and the body are covered with scales or modified hairs, which give moths and butterflies their resplendent colors. The mouth-parts of adults often form a sucking tube that is rolled into a tight coil when not in use. Lepidoptera have four stages of development: egg, larva (caterpillar), pupa (cocoon or chrysalis), and adult. Most butterfly eggs are laid singly or a few at a time and are unprotected. A few moths lay a large number of eggs in one place and cover the mass with a protective coating which includes hairs and scales from the female's body. Caterpillars of both have

scales of butterflies

scales of moth

GIANT SWALLOWTAIL
caterpillar

chewing mouth-parts, primarily for feeding on plants. A handful of moth caterpillars sometimes do tremendous damage to crops. Most caterpillars have 6 true legs on the thorax, and from 4 to 10 unjointed false legs on the abdomen. A few have irritating hairs or stinging spines. Many caterpillars spin a silken cocoon, sometimes covered with hairs, in which they pupate. A true butterfly makes no cocoon but forms a chrysalis. Some Lepidoptera winter as pupae, others as eggs or caterpillars, and a few as adults. Adult moths are usually quite different from adult butterflies, though one group, the skippers, shares characteristics with both. Butterflies usually fly by day; moths mostly, but not exclusively, fly by night. Butterflies customarily rest with their wings held together over their back; most moths rest with their wings in a horizontal position. The antennae of butterflies are thin, ending in a knob. Those of moths rarely end in knobs and are often feathery. Though butterflies are more uniformly attractive, the moths form a larger, more diverse, and more important group.

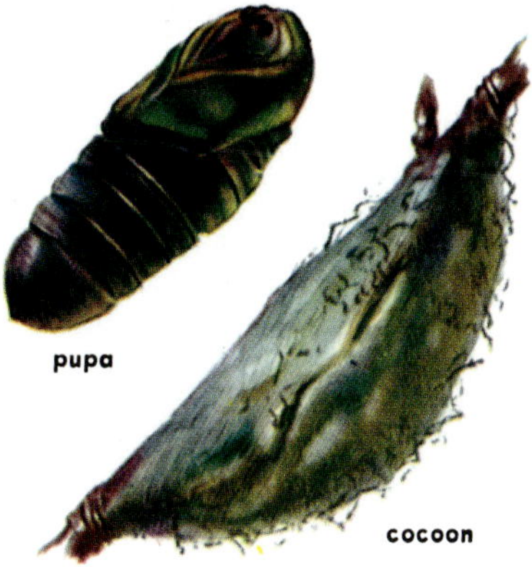

pupa

cocoon

MONARCH 1.3″ w. 4.0″

MONARCH This common and attractive butterfly is a migrant, arriving each year in the East from central Mexico and in the West from California. Males are identified by a black scent pocket on the third vein of the hindwing. Two or three generations are produced in one summer. In fall, swarms of adults migrate to California and Mexico, covering entire trees when they stop to rest. Adults taste bad to birds, the result of feeding as caterpillars on milkweeds, which contain natural poisons.

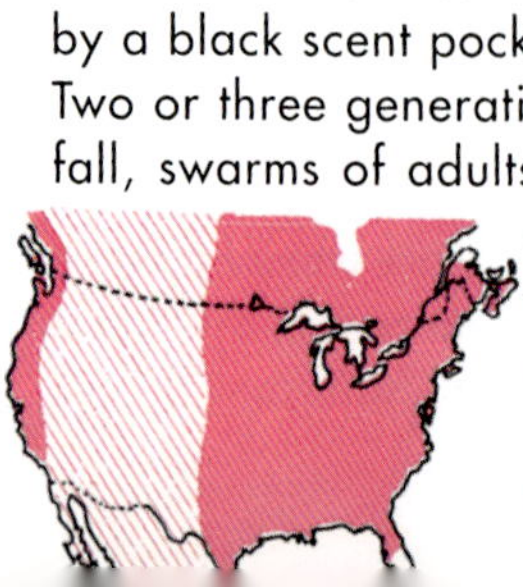

VICEROY
1.0″ w. 2.8″

VICEROY This butterfly resembles the monarch in wing color, pattern, flight, and some habits, but is smaller. The viceroy also has curved black lines crossing the veins of the hindwings and only one row of white dots along the edge of the hindwings. Its eggs, and the larvae which feed on poplar and willow, resemble those of the purples (p. 88), to which the viceroy is related. Partly grown larvae, less than an inch in length, hibernate in rolled leaves for the winter. There are two generations or more a year.

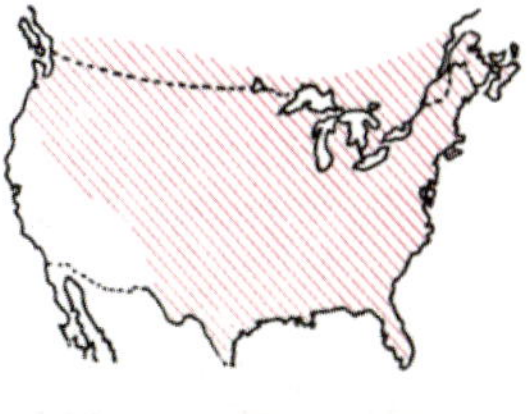

PURPLES The banded purple has a conspicuous white band across its wings with a border of red and blue spots on the hindwings. Eggs are laid on leaves of willow, birch, and poplar, on which the larvae feed. The red-spotted purple is named for its red spots on the underside along the wing borders and at the base of the hindwings. Its larvae feed on wild cherry, willow, and other trees. These two butterflies are geographic races of a single species.

COMMON BUCKEYE Though usually found in the South and West, the buckeye occasionally migrates North. Eyespots make identification easy. It lays its green, ribbed eggs chiefly on plantain, snapdragon, and ruellia, plants on which its larvae feed. The bases of the larval spines have a distinct blue-violet cast. The larva forms a brown chrysalis from which the adult eventually emerges. Two similar species in Texas and Florida have smaller eyespots.

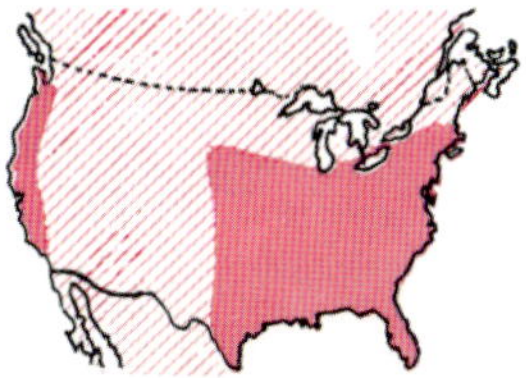

GREAT SPANGLED
FRITILLARY
1.0″ w. 3.5″

MEADOW
FRITILLARY
0.6″ w. 1.7″

VARIEGATED
FRITILLARY
larva

REGAL
FRITILLARY
1.0″
w. 3.4″

upper side

GULF FRITILLARY
0.9″ w. 2.9″
SILVER-
BORDERED
FRITILLARY
0.6″ w. 1.5″
GULF FRITILLARY larva
REGAL
FRITILLARY
under
side

FRITILLARIES One of the largest groups of butterflies, the fritillaries are found not only in this country but in many parts of the world. They are members of a family which uses only 4 of their 6 legs for walking. The front legs are reduced in size and held close to the body. Fritillaries are mostly medium-sized butterflies, orange or reddish above, with silvery or light spots on the underside of the hindwings. Sometimes the males are a brighter orange on top than the females.

The eggs are generally barrel-shaped and ornamented with ridges. All the caterpillars are spiny, with the slightly longer spines on the head. Many of our species overwinter as a tiny unfed larva. Most feed at night on plants such as violets. While the early stages of the common fritillaries are known, there are many less common species, and information on the egg and caterpillar of these species is incomplete. The chrysalis of fritillaries is usually angular, forked at the top, and bordered with knobs. It is often brownish. Of the many fritillaries, those illustrated are among the most common and best known. The Gulf fritillary is a longwing butterfly, not a "true" fritillary.

GREAT SPANGLED FRITILLARY larva

CHECKERSPOTS are a group of medium-sized, festively colored butterflies. The Baltimore is common locally in June and early July in wet meadows where turtlehead grows. The caterpillar is black with orange bands and shiny black spines. The wing pattern of the anicia checkerspot varies considerably. The caterpillar is black with orange spots at the base of the spines. It feeds principally on painted cups and monkey flowers. Nearly mature larvae of checkerspots wander widely.

ANGLEWINGS The angular, notched forewings of this group of butterflies are easy to spot. The hindwings often have short tails. Of 10 species in this country, the question mark (also called the violet-tip), comma, and mourning cloak are best known. The brownish larvae of the question mark feed on hops, elm, and nettle. The comma caterpillar feeds on nettles, often forming a crude shelter by drawing the leaf edges together. Both caterpillars have branched spines.

MOURNING CLOAK
1.0" w. 2.6"

MOURNING CLOAK This butterfly, like other anglewings, hibernates as an adult and makes its appearance very early in spring, sometimes while there is still snow on the ground. Eggs are laid on twigs of poplar, elm, hackberry, or willow. The gregarious black, spiny caterpillars occasionally injure small trees by stripping the foliage. The mourning cloak is common and widely distributed over the entire Northern Hemisphere. In the North it has one brood a year; in the South, two.

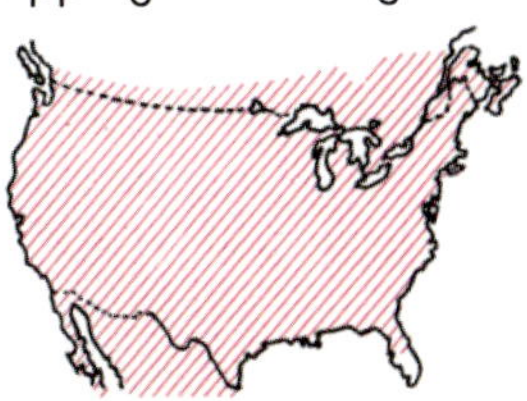

RED ADMIRAL, AMERICAN LADY, AND PAINTED LADY Two of these three closely related butterflies are difficult to distinguish. But the red admiral is clear and unmistakable because of the red bands on its forewings and red borders on its hindwings. The red admiral is found throughout the northern hemisphere. The caterpillar feeds on hops and nettles. Although similar to the painted lady, the underside of the hindwing of the American lady has two large eyespots. The American's larvae are spiny and black, with rows of white spots. It feeds on arrowweed,

cudweed, and other everlastings and frequently becomes a pest in flower gardens. The painted lady or thistle butterfly is reported to be the most widely distributed of all butterflies. This may be because the adults are migratory and very strong fliers. Also the plants its larvae eat, which include burdock, thistles, mallows, sunflowers, and nettles, are widely distributed. The color of the larvae varies greatly, ranging from yellow-green to almost black.

NYMPHS AND SATYRS About 50 species make up this large group of small to medium-sized butterflies, most of which are some shade of brown. Their characteristic markings are eyespots on the undersides of the wings, always positioned near the wing margins. Most prefer open woods and mountain areas, generally in the North. The green caterpillars of these species bear numerous short hairs, and taper toward a forked tail. The larvae of the eyed brown have a pair of red horns at each end of the body. Their food is primarily sedges and other grasses.

Pearly-eyes often feed at wounds in trees and are active late in the day. The common wood nymph has a pair of conspicuous eyespots on each forewing. The larvae lack the long horns of the first two species. The little wood satyr is smaller and has eyespots on both the fore- and hindwings that are marked with pairs of white dots.

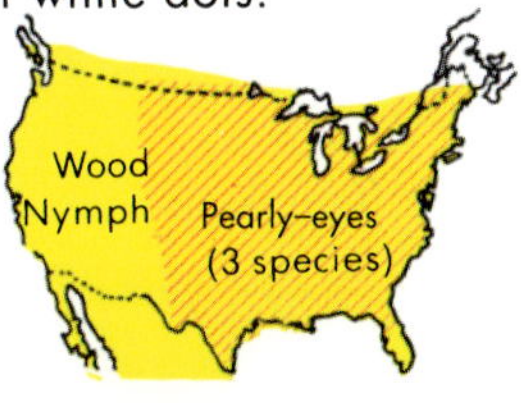

PURPLISH COPPER
0.5″ w. 1.2″

GRAY HAIRSTREAK
0.4″ w. 1.2″

AMERICAN COPPER
0.4″ w. 1.0″

BRONZE COPPER
0.5″ w. 1.5″

HAIRSTREAKS, COPPERS, AND BLUES This large family of small butterflies includes over 2,000 species. Not many species are found in this country, but these butterflies are common. The family has three groups. The hairstreaks are usually brownish or bluish, with hair-like tails at the tip of the hindwings. About 90 species of hairstreaks live in the United States. The second group, the coppers, are nearly all a copper-red color with black markings. There are some 16 species. The American copper is probably the most common. This brilliantly orange butterfly is highly territorial. It is found everywhere east of

EASTERN TAILED–BLUE
0.4″
w. 1.0″

SPRING AZURE
0.4″ w. 1.1″

under

MARINE BLUE
0.4″ w. 1.1″

upper

WESTERN PYGMY BLUE
0.2″ w. 0.6″

the Rockies. The last group, the blues, are very small. Western pygmy blue is our smallest butterfly. Blues are more common in the West. At least 35 species, variable in form and difficult to identify, are listed for the United States. The spring azure has over 13 different variations. The caterpillars of this family are short, thick, and slug-like and covered with fine hair. They feed on a variety of plants, principally legumes, but also oak, hickory, hops, and sorrel. One species, the harvester, has a carnivorous caterpillar that devours aphids.

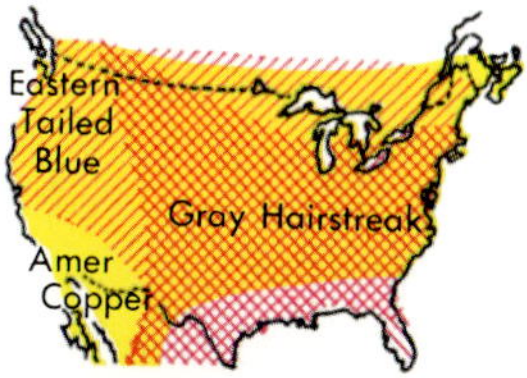

CABBAGE WHITE
0.7″ w. 1.8″

CABBAGE WHITE This all-too-common species ranges over most of the Northern Hemisphere. It first entered the eastern part of this country in 1868 and within 20 years had spread to the Rockies. Now these insects are found in every cabbage field; the green caterpillars feed also on mustard, nasturtium, and related wild plants. They are often a pest in gardens. Two or three broods mature each year, the last brood hibernating as the chrysalis and emerging in early spring.

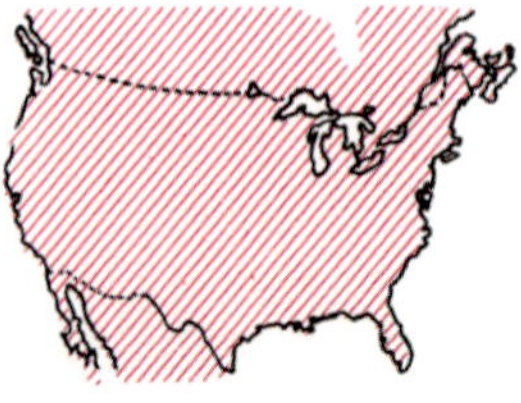

SULPHURS Dozens of common or clouded sulphurs may be seen clustered together over roadside puddles. Their color varies: Females are paler, with yellow spots in the borders of their forewings. The orange scales on the alfalfa butterfly may be reduced to a patch on the forewing. The alfalfa is common along roadsides and in fields and gardens. Its larva is similar to the common sulphur's. The larvae of both butterflies feed on clover, alfalfa, and other legumes.

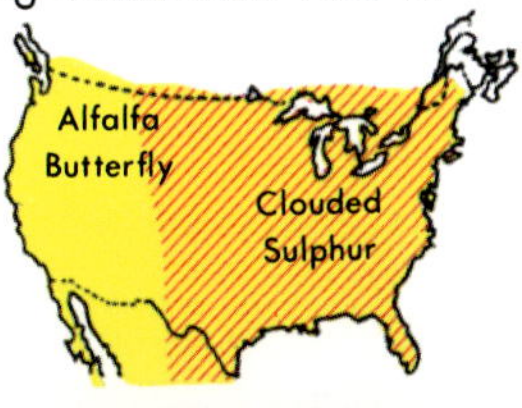

SPICEBUSH SWALLOWTAIL
0.8″ w. 3.5″

PHOEBUS PARNASSIAN
0.8″ w. 2.4″

GIANT SWALLOWTAIL
1.0″ w. 4.4″

ZEBRA SWALLOWTAIL
0.8″ w. 2.7″

BLACK SWALLOWTAIL 0.8″ w. 2.8″

TIGER SWALLOWTAIL
0.9″ w. 3.8″

SWALLOWTAILS These are our largest and most familiar insects. Over 30 species occur in the United States and many others are found elsewhere, making the swallowtails a group that is widely known, admired, collected, and studied. Closely related to the swallowtails are the parnassians of the West. These primitive swallowtails lack the "tail" on the hindwings. The caterpillars pupate on the ground. The Phoebus Parnassian (p. 104) is an example of this group.

Swallowtails are predominantly black or yellow. Some species occur in several forms. Female tiger swallowtails may be either yellow or black, the black form being more common in the South. Eggs are smooth, round, and flattened at the base. The caterpillars, shown with their food plants on the next page, are generally smooth. They lack spines, although pipevine swallowtails have numerous fleshy tentacles (p. 105). Larvae have a reversible organ behind the head, which emits a repulsive odor that protects them from some predators. Their green or brown chrysalis rests on its end, supported by a loop of silk at the middle.

The swallowtails illustrated here are easy to identify. The giant swallowtail is the largest, with a wingspread of 4 to 5 ½ inches. It is most common in the Southeast. Its larvae are occasionally destructive to citrus orchards. Both male and female black swallowtails have black wings marked by a band of prominent yellow spots. A second band of smaller yellow spots line the margins of its hindwings. The spicebush swallowtail has greenish spots on the margins of its hindwings. The pipevine, with its metallic blue scaling, has no yellow or green spots on the tops of its wings.

SPICEBUSH on sassafras

ZEBRA on pawpaw

PIPEVINE on pipevine

BLACK on parsley

GIANT on orange

TIGER on wild cherry

SWALLOWTAIL CATERPILLARS
and their food plants

SKIPPERS About 275 kinds of skippers, or about one-tenth of the total number found in the world, are native to the U.S. and Canada. Their rapid, darting flight gives them their name. These small butterflies share characteristics with moths. Some rest with their hindwings or both wings horizontal as moths do. Their smooth caterpillars have large heads and thin "necks." They feed on grass, legumes, and other plants. The silver-spotted skipper is the most common of the larger skippers. Many are much smaller.

AILANTHUS SILKMOTH
1.0″ w. 4.2″

AILANTHUS SILKMOTH This moth is native to China, where a coarse grade of silk is obtained from its cocoons. It was imported to Philadelphia in the nineteenth century to start a silk industry, but the project failed. Some moths were released, and by 1861 the species had become established. It feeds chiefly on ailanthus trees and is not harmful. Its numbers have declined, and it has disappeared from several cities where it formerly was found. It is our only large moth with white tufts on the abdomen.

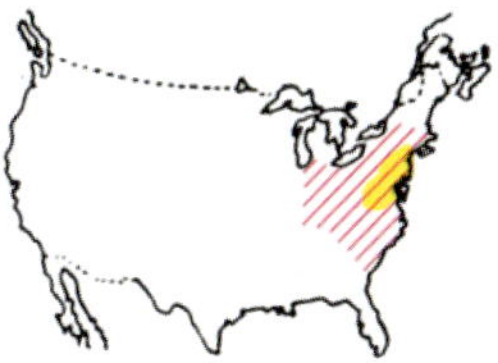

CECROPIA MOTH The enormous larvae of the cecropia feed on cherry, maple, willow, ash, lilac, and many other plants. The large, tough, brown cocoons are firmly attached to branches and are easily found in winter. Outdoors, the huge moths emerge in late spring or summer, but when cocoons are brought indoors they hatch earlier. The emergence of the adult is an amazing sight. The wrinkled, velvety wings unfold and expand until they are 5 or 6 inches across.

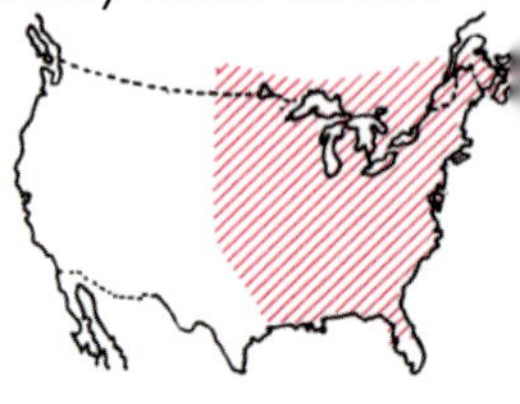

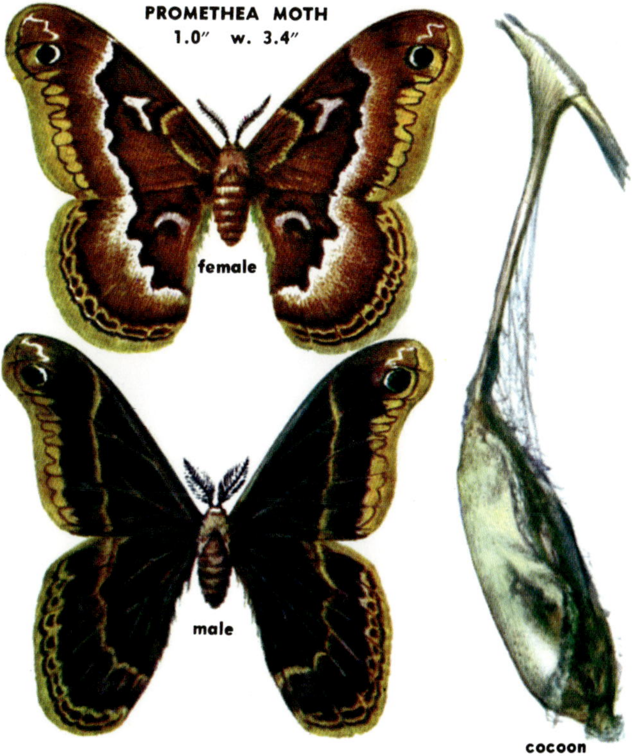

PROMETHEA MOTH This moth, sometimes called the spicebush silkmoth, has a bluish-green larva with two pairs of short red horns near its head. It feeds on many plants including sassafras, wild cherry, tulip tree, sweet gum, lilac, and spicebush. The compact cocoon is wrapped in a dry leaf and attached to a twig, where it will dangle all winter long. Adults emerge in spring. The male can be recognized by his deep chocolate brown color. The female is a bit larger, lighter, and redder, with slightly different markings.

POLYPHEMUS MOTH
1.3″ w. 5.3″

POLYPHEMUS MOTH Because of its gigantic size and the eyespots on the wings, this night flying silk moth was named after the one-eyed giant, Polyphemus, of Greek mythology. The green larvae, sometimes over 3 inches long, feed on oak, hickory, elm, maple, birch, and other trees and shrubs. They spin their broadly oval cocoons either on the ground or attached to a twig. There is a single brood in New England and up to three in the South.

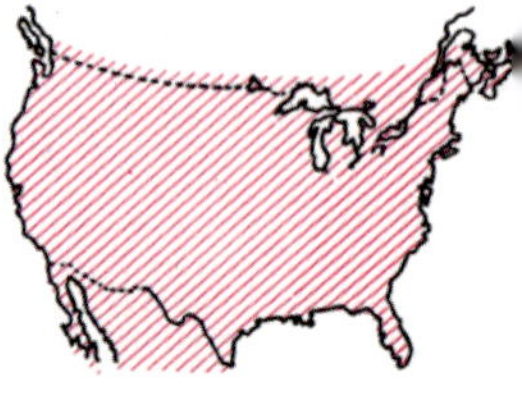

IO MOTH female
1.1″ w. 2.8″

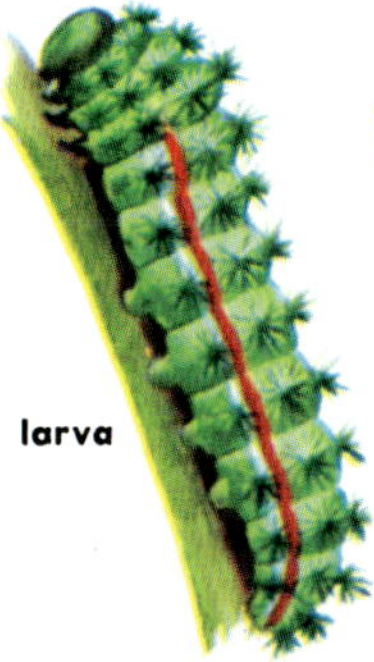

larva

male
0.9″ w. 2.5″

IO MOTHS Three very similar species are found in this country. Larvae ae easily recognized by their horizontal red and white stripe. They are fond on willow and other wild plants, and on garden plants such as corn. Their sharp spines are mildly poisonous, but Ichneumon wasps (p. 137) often attack the larvae. The cocoon is spun on the ground in dead leaves. The adult male is smaller than the female and has bright yellow fore-wings.

LUNA MOTH This handsome moth, with its striking long tails and delicate green color, makes a lasting impression on those who see it for the first time. It is a favorite with collectors. The large larvae feed on many trees, including sweet gum, walnut, hickory, birch, and persimmon. The cocoons, though similar to those of the polyphemus moth, are lighter and more papery. Some adults have more purple on the borders of their wings than others. One brood each year in the North; up to three in the South.

IMPERIAL and REGAL MOTHS These moths are closely related to the large silk moths. The hairy imperial moth caterpillar has short horns near the head. It varies from pink and green to brown or charcoal. It feeds on pine, hickory, oak, maple, and other trees. No cocoon is formed—the pupa rests in a cell in the ground. The larva of the regal moth has large, red and black curved horns. It feeds on walnuts, sweet gum, hickory, and sumac.

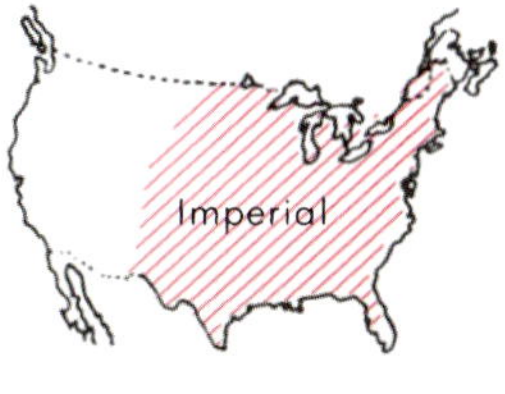

SPHINX MOTHS More than 120 species of these thick-bodied, narrow-winged moths live in this country. Their common names, tomato worm and tobacco worm, indicate the food plants sought by the larvae of a few species. Some species feed on potatoes; others eat birch, poplar, willow, grape, cherry, and other plants. They eat leaves and sometimes the fruit. The larvae are large and usually have a tail or horn. Some rear back in a snake-like fashion and will thrash and regurgitate if disturbed, but they are harmless. Braconid wasps lay their eggs inside living sphinx moth caterpillars. The wasp larvae feed on the

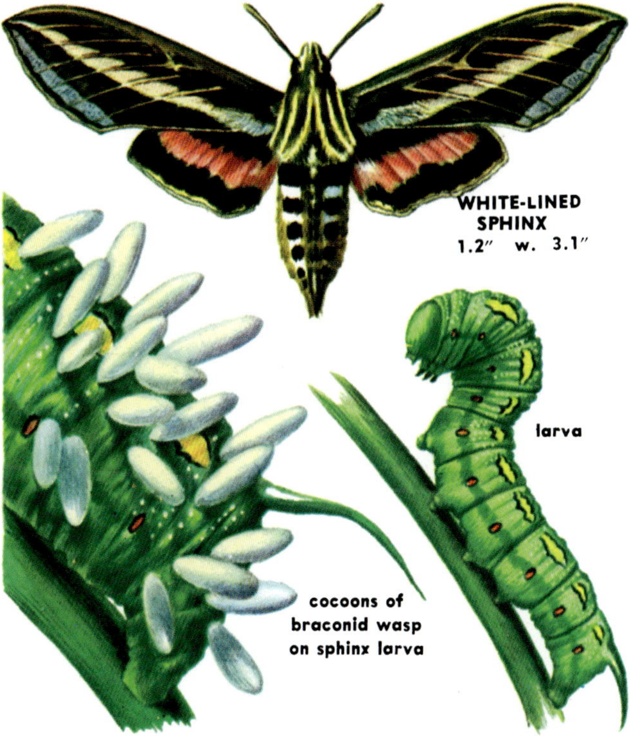

caterpillar and then tunnel up and out and form a cocoon. Caterpillars covered with dozens, even hundreds, of wasp cocoons are often seen. The caterpillar is always mortally wounded. Sphinx moth caterpillars pupate in the ground, and some can be recognized by the loop at one end. An adult sphinx moth's mouth is a long sucking-tube, which enables it to reach nectar and to pollinate tubular flowers such as nicotina, petunia, honeysuckle, trumpet vine, and many others.

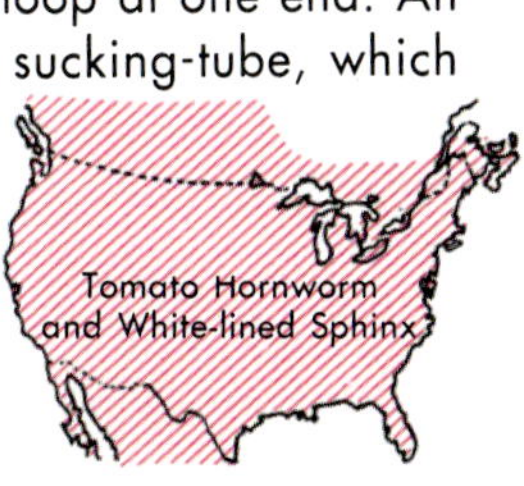

CLOUDED LOCUST UNDERWING 0.8″ w. 2.4″

UNDERWING MOTHS There are over 110 species of these attractive moths in the United States, and a variety can be found in every region. They can be attracted to a tree trunk or stump by painting it with a mixture of brown sugar, beer, and fermented fruit. When an adult rests on bark with its wings folded, it can scarcely be seen. In flight, the bright colors of the underwings make a sharp contrast to the drab pattern of the forewings.

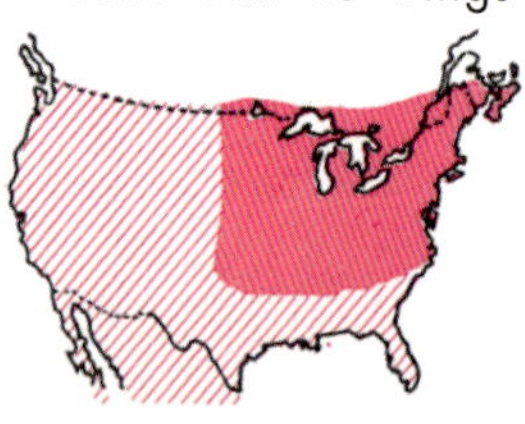

CORN EARWORM AND BORERS The greenish or brown larva of the corn earworm feeds on corn and other garden crops. It pupates underground. The European corn borer became established after the turn of the century and has spread widely since, damaging much corn as it spread. The larvae bore into stalks, weakening and breaking them. It is considered one of the most serious pests of corn. Larvae also damage asters, clover, potato, and other plants.

SALT MARSH TIGER MOTH AND ISABELLA MOTHS represent a family of some 265 American species. The Salt Marsh Tiger Moth is easily identified by its spotted abdomen. The female's hindwings are white. Caterpillars feed on many plants, including garden species. Occasionally they will catch other caterpillars and consume them. The banded woollybear, the well-known Isabella moth caterpillar, is abundant in fall, feeding on asters, sunflowers, clover, and other plants. The larvae of both species mix body hair with silk in making their cocoons.

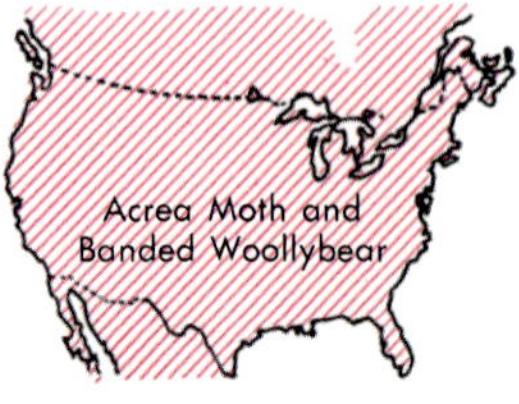

WHITE–MARKED TUSSOCK MOTH
and larva on maple

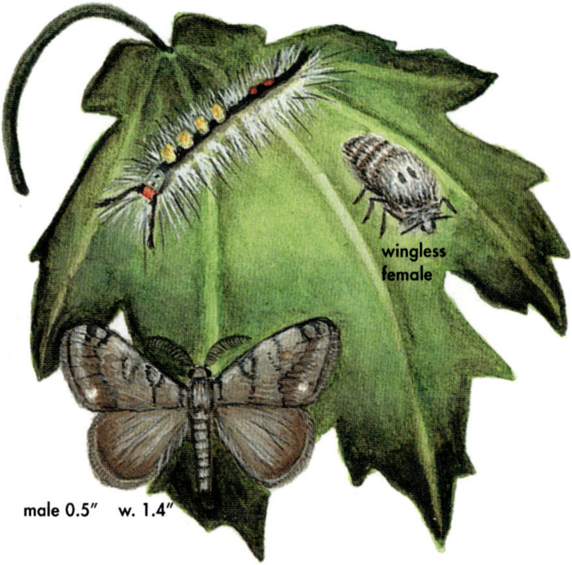

WHITE-MARKED TUSSOCK MOTH The larvae, often seen in late summer, are recognized by their tufted white hairs and attractive contrasting colors. The bright, red spots on the abdomen are defensive glands that produce a mixture of chemicals. Some people ae highly allergic to these caterpillars. The wingless females lay egg masses on the surface of their cocoons. Eggs hatch in spring, but the larvae are not noticeable until summer. The drab adults are seldom noticed.

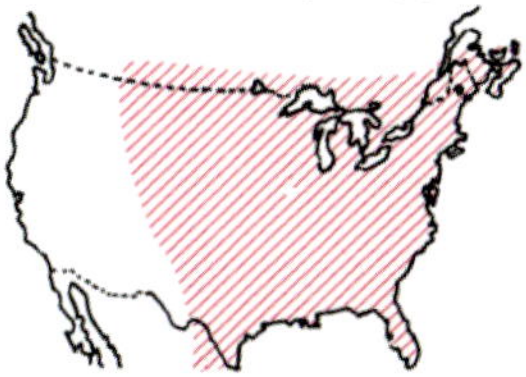

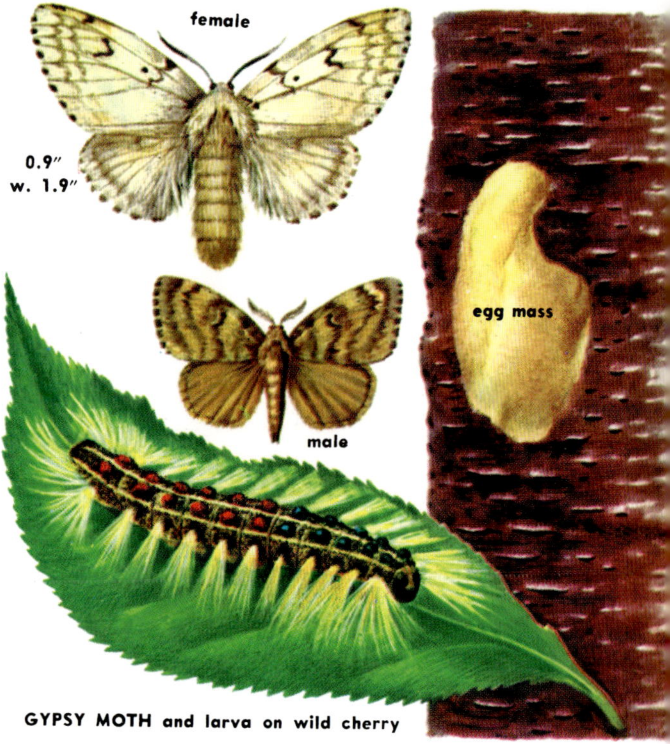

GYPSY MOTH and larva on wild cherry

GYPSY MOTH The gypsy moth is a European relative of the white-marked tussock moth (p. 121) which, unfortunately, has become established in this country. Touching the larvae of gypsy moths can cause an allergic reaction in some people. Larvae feed at night, often in great numbers, eating the leaves of many trees. Outbreaks can be

severe, with thousands of acres being defoliated. Oak forests are especially susceptible. Larvae pupate in midsummer and adults emerge shortly. The female seldom flies. She lays a mass of eggs that hatch in spring.

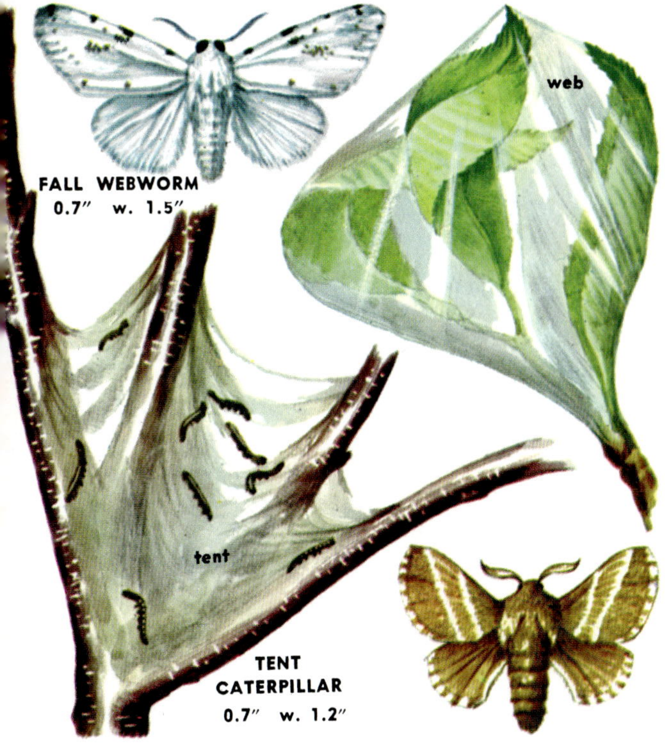

TENT CATERPILLARS AND FALL WEBWORMS

These pests of many forest, shade, and ornamental trees are not closely related, but are often confused. Both build webs, but that of the fall webworm covers the leaves. The tent caterpillar web, built only in spring, is spun at the crotch of branches. Adult webworms, marked with variable black spots on forewings, lay eggs on leaves. Tent caterpillar egg masses are found on twigs in winter. Remove them to control this insect.

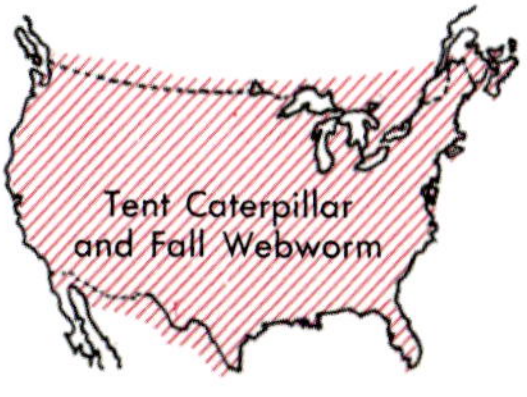

CANKERWORMS or inchworms are interesting to watch as they crawl along, curling and uncurling. Sometimes they spin a silken thread and hang suspended in mid-air, especially at night. Gall and spring cankerworms are occasional pests of apple and other fruit and shade trees. The wingless females lay their eggs on bark. After feasting on young spring foliage, the larvae pupate underground. The adults emerge late in fall and can be trapped by placing bands of sticky paper around tree trunks.

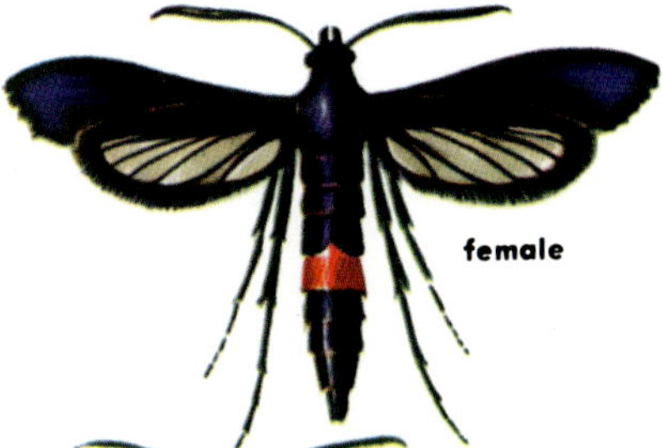

female

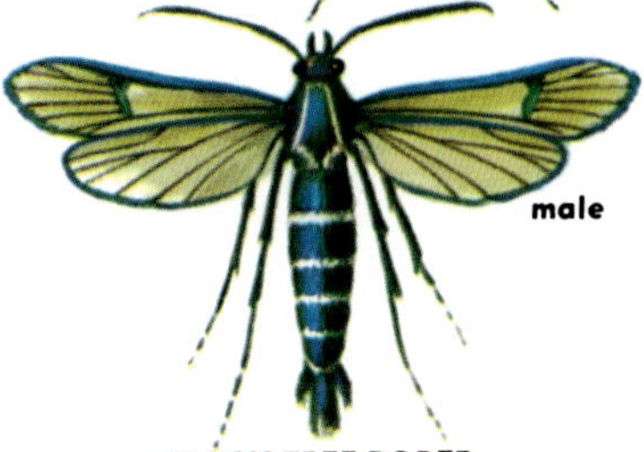

male

PEACH TREE BORER
0.6" w. 1.1"

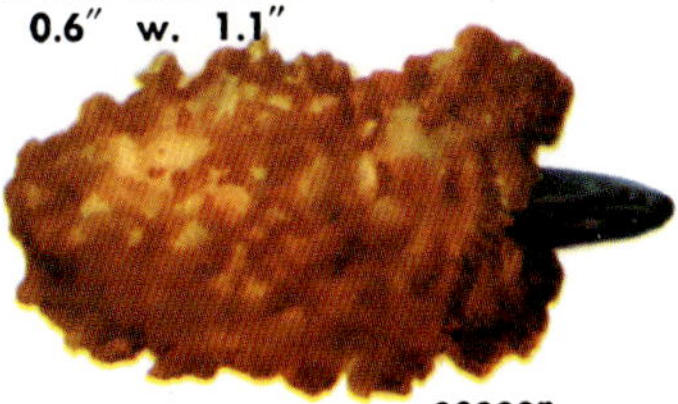

cocoon

gum on trunk

PEACH TREE BORER lays its eggs in fissures in a tree's bark. The larvae bore into the tree as soon as they hatch. Their burrows can be recognized by oozing sap on the bark surface. After spending winter as a larva, the borer pupates in a crude cocoon. The adult emerges in about a month and mates. The female lays a new mass of several hundred eggs. This pest is believed to have fed on wild plum and cherry before peaches were introduced.

CODLING MOTH This pest, introduced from Europe, is now found across much of North America. It deposits tiny eggs on leaves. As the eggs hatch, the larvae locate and bore into the new fruit of apple, cherry, peach, pear, and walnut. Mature larvae leave the fruit to pupate on the bark. There may be several generations a year. Fully grown larvae hibernate and pupate the following spring after additional feeding.

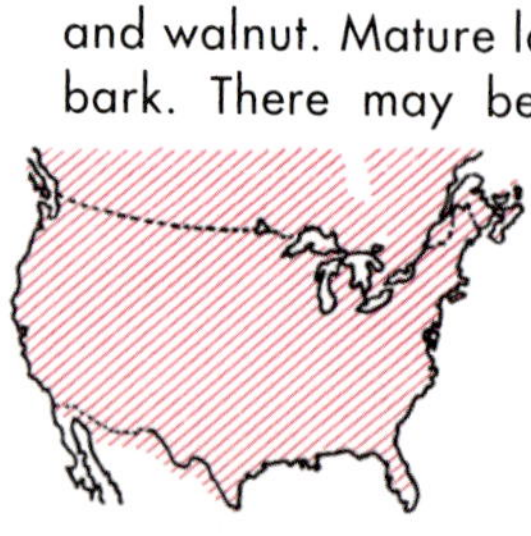

BAGWORMS have a strange life history. The eggs hatch in spring and the young larvae feed on the leaves of many kinds of plants. They build their bags from bits of leaves and twigs as they feed. Eventually they bind their bags to twigs and pupate. The adult male emerges, seeks a female, and mates. After mating, the wingless and legless female crawls back into her "bag" and lays hundreds of yellow eggs. Several related moths also make bags, each with a design characteristic to that species.

SCORPIONFLIES are harmless, but resemble scorpions because of their extraordinary reproductive structures. Eggs are deposited in the soil, where they develop into larvae. The larvae change into pupae, from which the fly-like adults emerge. They live as scavengers, feeding on dead or disabled insects. Adults are often seen perched on plants. They are not strong fliers. One group of minute, almost wingless, scorpionflies lives in northern woods and is active even on snow in winter and early spring.

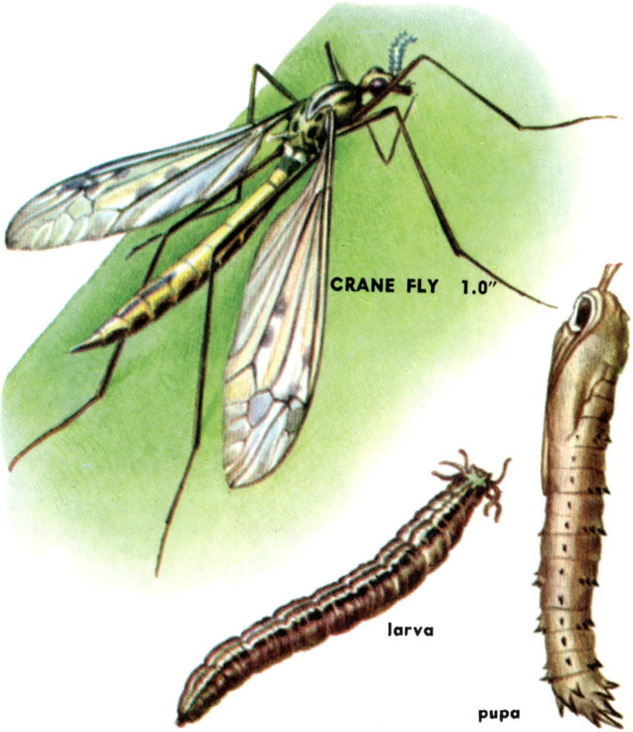

CRANE FLIES Often seen around lights, adult crane flies are sometimes mistaken for giant mosquitoes, but they do not bite. There are more than 1,600 species in North America. Some are apparently predators, but in many species the adults do not eat at all. The female lays several hundred eggs on damp soil or in water. The larvae burrow into the ground and eat decaying wood. Only a few attack plants. After several weeks they pupate. Adults appear about a week later.

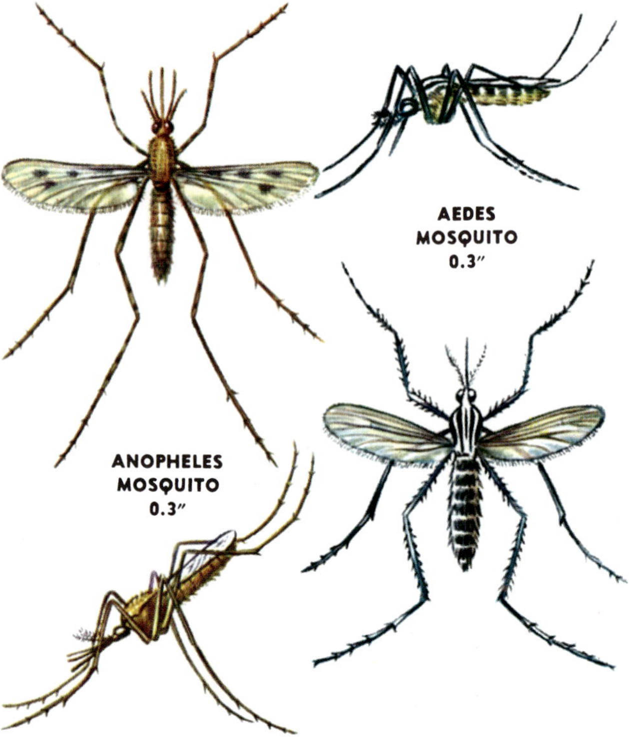

MOSQUITOES This large group of small but important insects has been widely studied as part of public-health campaigns against malaria and yellow fever. The local and regional conquest of malaria is a scientific milestone. The common carriers of the disease, the Anopheles mosquitoes, are recognized by the "three-pronged" beak of the female and by the tilted position they assume when resting. The Aedes mosquitoes, one of which carries yellow fever, look more like the common house and swamp mosquito. The disease is now limited to tropics and sub-

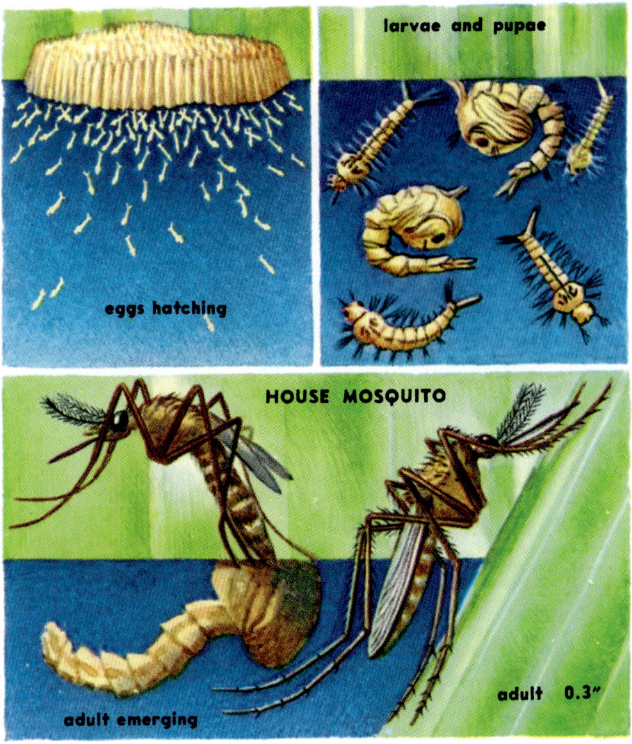

tropics. Female Culex mosquitoes lay rafts of several hundred eggs, which hatch in a few days into larval wrigglers. In a week or so these pupate, and the adults soon emerge. Male mosquitoes feed on nectar. Females bite to feed on blood and can transmit disease. Historically, mosquitoes have been controlled by draining swamps and by the wide use of insecticides. The ecological consequences associated with these methods have not been trivial. Both as adults and larvae, mosquitoes are important food for animals such as fish, birds, and dragonflies.

FLIES include serious pests of plants and animals, but the vast majority are harmless and many are beneficial as pollinators, parasites of pests, and food for other animals. March flies (which are more common in late spring) are often seen on flowers. Robber flies prey on many insects, some larger than themselves. Large numbers of deer flies can make hiking and camping miserable. The horse fly, sometimes a full inch long, can give a severe bite. Large horse flies feed principally on large animals such as deer and, of course, horses. Some flies transmit disease when they bite infected animals and then bite others.

male 0.2″ beaded-wing vestigial-wing

FRUIT FLIES These small and important flies (several hundred species) are often seen around rotting or fermenting fruit and fungi. Their claim to fame rests on scientific uses to which they have been put. They are easily grown in the laboratory. One species, in particular, has been used in studies of inheritance. The fact that their life cycle is less than 2 weeks enhances their value in this work.

larva

BLUEBOTTLE FLY 0.5″

GREENBOTTLE FLY 0.5″

pupa

BLUEBOTTLE AND GREENBOTTLE FLIES are attractive insects whose larvae (maggots) consume and recycle decaying or dead organic matter. Eggs are laid on dead animals, garbage, sewage, or in open wounds of animals. Some related species parasitize and kill animals and even man. Eggs hatch very soon after being laid; the larvae may mature in less than 2 weeks. The short life cycle allows for several generations each season.

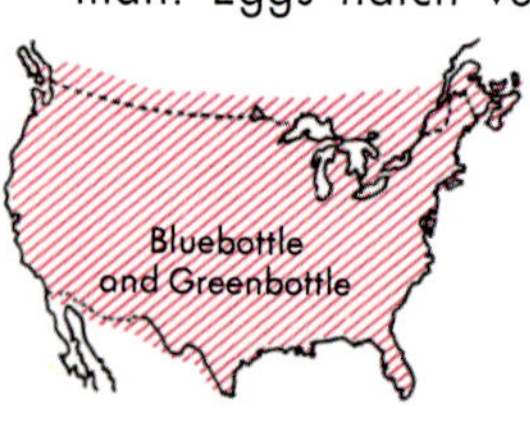

TACHINID AND OTHER FLIES Tachinid flies are thought of as beneficial insects because many help control injurious ones. About 1,360 species have been reported in North America. Syrphid flies, known as flower or hover flies, are handsome insects. They are often seen approaching a flower, coming to an abrupt stop, and hovering in mid-air. A few have predatory larvae that eat aphids and scales. Bee flies are fuzzy, handsomely marked insects. They can be wary and difficult to approach. Their larvae feed on the larvae of bees and other insects. They are especially common in sandy areas.

HORNTAILS AND SAWFLIES are ancient wasps. They can be distinguised by the shape of their body, which is not pinched like a bee's. Sawflies are abundant, leaf-eating insects. Many species are gregarious; a few are pests. Horntails lay their eggs in dead or dying trees. Their larvae are borers that tunnel deep into wood. Horntails are hosts to the ichneumons (next page), which parasitize the larvae.

ICHNEUMON 1.4″

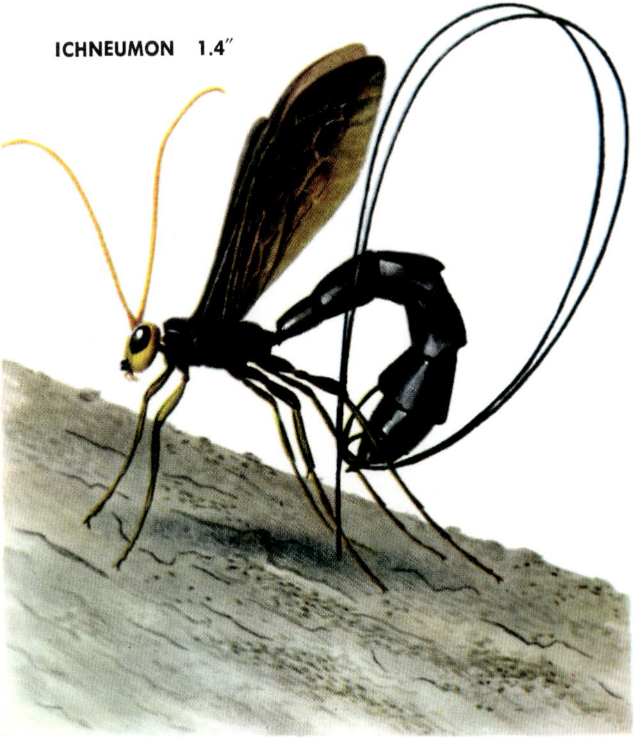

ICHNEUMONS probably number 70,000 species worldwide, over 3,000 of which are American. They belong to the same order of insects as bees and wasps and play an important role in controlling many harmful insects. Their larvae are parasites of caterpillars, beetles, flies, and other pests. Amazingly, the hair-thin ovipositor of the female long-tailed ichneumon can be drilled into several inches of wood to lay her eggs on horntail larvae.

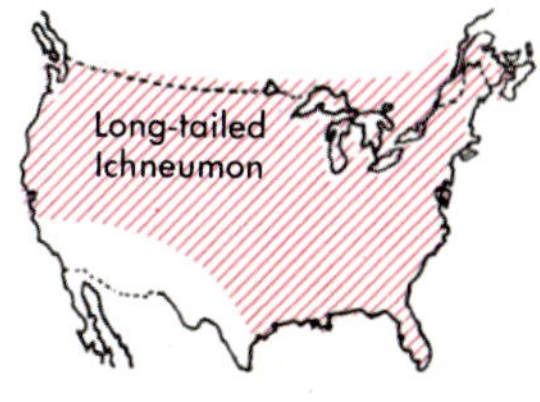

MUD WASPS Several families of wasps are represented among those building their nests with mud. Potter wasps are solitary, each building a vase-shaped nest of mud on plants. These wasps provision their nests with caterpillars and beetle larvae. Most mason wasps, in the same family as the potters, nest in burrows in the soil, but the species illustrated makes a clay nest on a branch. Best-known mud wasps are the mud daubers, which make large nests on walls in attics or deserted buildings. The female builds the nest of many mud cells. She places in each nest

BLACK–AND–YELLOW MUD DAUBER 1.0″
and nest

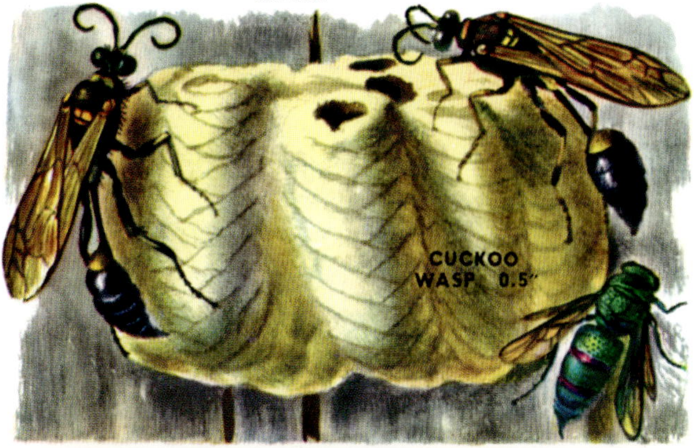

several paralyzed spiders or insects (that she has stung) before she lays the egg and seals the cell. The blue mud dauber uses nests made by the black-and-yellow mud dauber. The cuckoo wasp, named after the European cuckoo, awaits its opportunity and lays its eggs in the nest of a mud dauber or other wasp while the latter is off searching for a victim. The cuckoo wasp larvae feed on the spiders and other prey that had been provided for the young mud dauber.

INSECT GALLS are lumps formed when flies, moths, or small, wasp-like insects (Cynipids) lay eggs in plant tissues. Each insect selects a specific plant. As eggs develop and hatch, the plant tissues around the larvae begin to swell, forming a characteristic gall. The larvae feed on plant tissues and juices and usually pupate in the gall. Some galls are large and woody, some soft, some knobby and spiny. Best-known galls are the oak apples and the galls commonly seen on rose, blackberry, and goldenrod.

CARPENTER ANTS Of over 2,500 species of ants known, all are social animals, living and working together in ways that have astonished laymen and naturalists alike. Among the most familiar of insects, they have inspired many a comparison with human society. Carpenter ants build nests and burrows in dead wood, logs, and the timbers of buildings. They also sometimes damage old or rotten timbers. They are found in temperate regions all over the world. The workers, who are infertile females, are among the largest ants in our region.

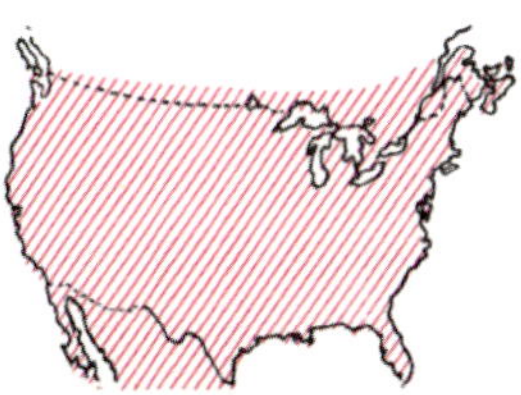

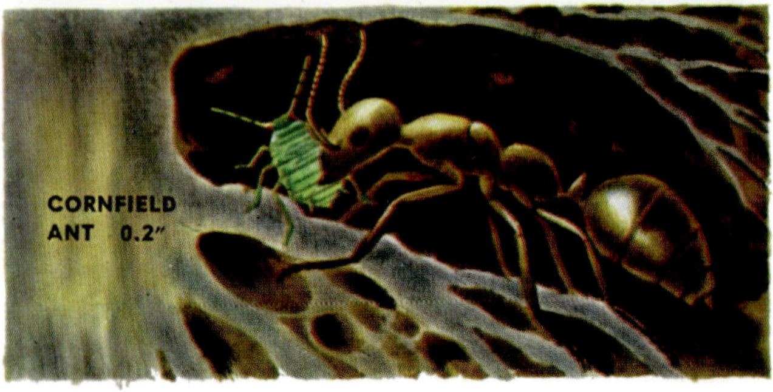

FIRE AND CORNFIELD ANTS Despite their small size, fire ants can inflict a painful sting. In the Southeast they have been known to attack baby birds and sting them to death. In humans, a small blister usually develops around each sting. The fire ant is a serious, invasive pest in yards, pastures, and other farmland. Cornfield ants are less ferocious and more interesting. These ants eat the sweet secretions of cornroot aphids. Aphids lay eggs in the ant burrows. When the eggs hatch in spring, the ants place the aphids on knotweed roots until the corn is planted and growing. Then the ants transfer the aphids to the corn roots, thus insuring a constant food supply.

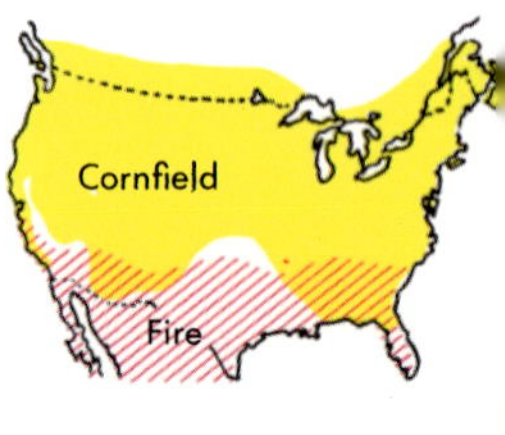

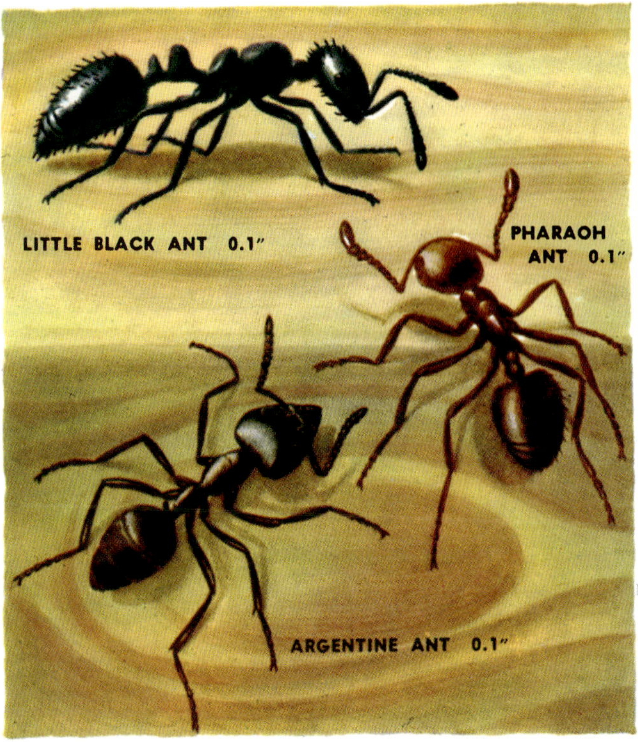

HOUSEHOLD ANTS Pharaoh ants are small but numerous invaders of homes, feeding on sweet foods. The Argentine ant, native to that country and Brazil, was first found in New Orleans in 1891, and has since become a serious household and garden pest in the South and California. Fortunately it is a semitropical species and its movement northward is limited by cold weather. Little black ants are usually found outdoors, for example, along sidewalks. These and other ants protect aphids from other insects. In return, the ants gather honeydew secreted by the aphids.

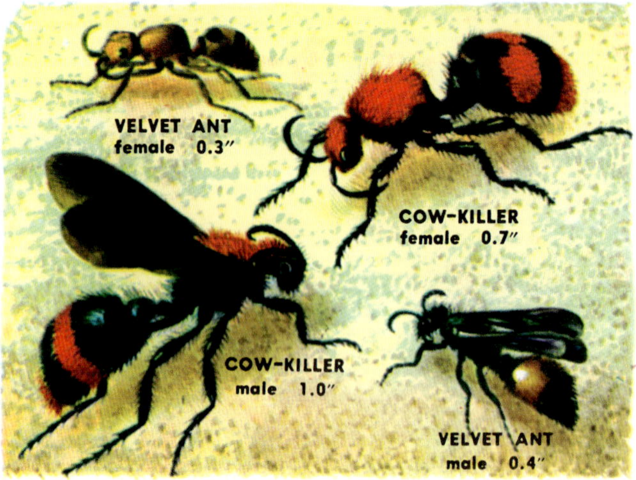

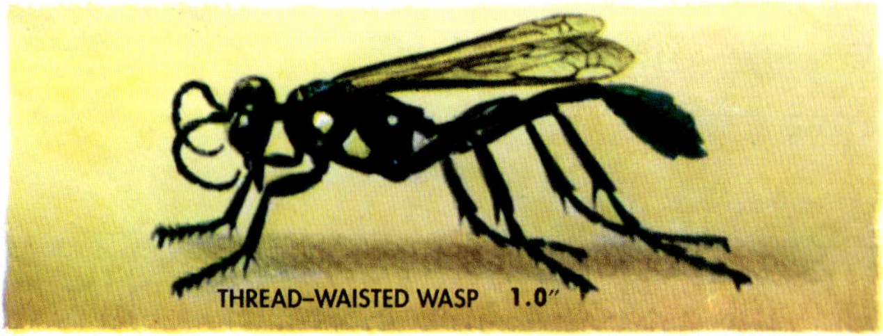

VELVET ANTS are given their name because the wingless female looks like an ant, but they are actually hairy wasps. They are parasites of other wasps and bees, especially species that live alone. A female velvet ant crawls down into a burrow and kills the owner with a powerful sting. She then lays her egg on the owner's larva, which her own larva later eats. The male velvet ant does not sting. The thread-waisted wasp, illustrated above, is one of several kinds parasitized by velvet ants.

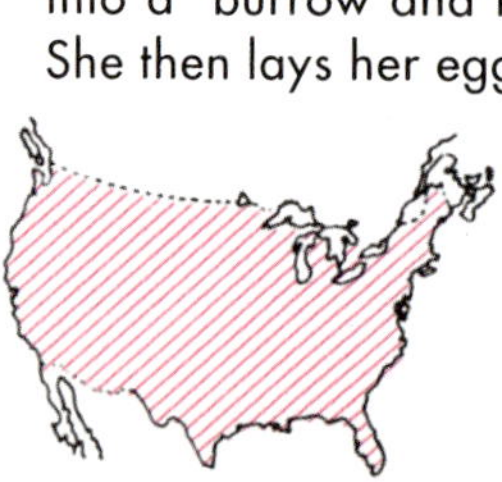

CICADA KILLER 1.6″

CICADA KILLER This large solitary wasp provides food for her offspring by paralyzing cicadas with her sting and storing them in her burrow. Because adult cicadas are heavy, she may drag her prey up a tree to get enough altitude to fly it to her burrow. She places the cicada in a side passage and lays an egg on it. When the egg hatches, the larva feeds on the helpless cicada. The larva pupates in a loose cocoon.

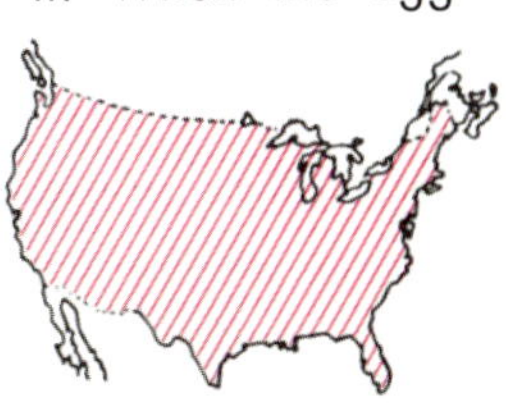

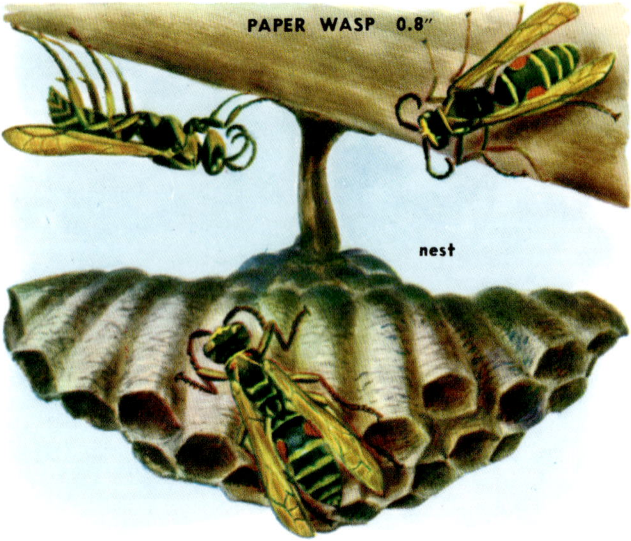

PAPER WASPS are the common wasps everyone learns to know sooner or later, often by painful experience. The common paper wasp builds an unprotected paper nest out of wood it has chewed up. The nest is hung under eaves or in any sheltered site. After the eggs hatch, the young are fed daily until they pupate. The bald-faced or white-faced hornet builds an oval, covered nest, some large enough to accommodate over 10,000 hornets. Unfertilized eggs develop into male drones. Fertilized eggs grow into workers or queens, depending on the amount of food provided. Queens make small brood nests in early spring. Closely related yellow jackets build nests of varying sizes, some underground or hidden in rock walls, under logs, in the walls of buildings, or even in an empty field mouse nest.

nest
BALD-FACED
HORNET
drone
0.7″
worker 0.6″
queen 1.0″
YELLOW JACKET
0.8″

BEES More than 3,300 species of bees found in North America. They are important pollinators of fruits and flowers and crops such as cotton and clover. Most are solitary, nesting in the ground or in natural cavities, but a few live in complex societies. Honey bees are probably the best known. A colony may contain up to 50,000 bees, including a fertile female (queen), infertile female workers, and male drones. Africanized honey bees are the result of interbreeding between honey bees and bees from Africa. They are called "killer bees" because they are easily dis-

turbed, and large numbers of bees often sting at one time. Bumblebees are among our largest bees. They are often the first insect seen in the morning and the last at night. Their nests are established underground each year by queens who have hibernated through winter. Sweat bees, small and often brilliantly colored, nest in the ground. They are attracted to the salt in perspiration, giving them their name. Leafcutting bees cut leaves from roses and other plants and line their underground nests with them.

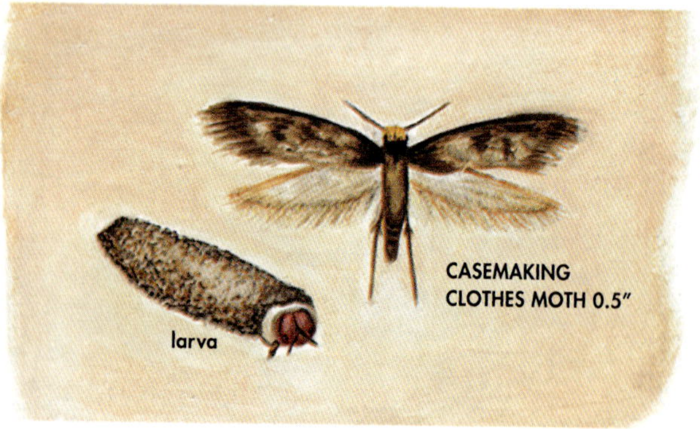

PESTS OF CLOTH AND CLOTHING Carpet beetles eat all kinds of animal matter, thriving on rugs, wool, fur, leather, and hair. The larvae do the damage. They can be controlled by repellents and poisons, but these toxic chemicals should be used with caution. Female clothes moths lay white oval eggs on clothing, leather, or cloth. These soon hatch into larvae and soon build a purse-like case from which they feed.

PESTS OF ANIMALS are many. Fleas found on dogs are known to bite humans, and rodent fleas may, in some areas, carry plague. Fleas also infest other domestic animals. Larvae may live in bedding, but they can be controlled by vacuuming. The sheep ked, a species of fly, feeds by sucking blood. Its eggs hatch in the body of the female, and the larvae develop to maturity before being deposited. Control is achieved by shearing and dipping the sheep. The squalid duck louse is typical of bird lice (distantly related to species on page 52) which live on wild and domestic birds.

OTHER COMMON PESTS

Only a few of the insects in this book cause any harm to humans, but some other species can be bothersome pests. These include:

SILVERFISH Primitive, soft-bodied insects which eat starch from bookbindings, wallpaper, and clothing. They are common indoors, especially in warm places.

AMERICAN DOG TICK Not an insect: it has 4 pairs of legs. Common enough and often brought indoors after a walk through fields and woods. Check clothing and body immediately. If attached to skin, remove by touching tick with a swab dipped in alcohol. Ticks may transmit serious bacterial and viral diseases.

DEER TICK (*Ixodes scapularis*)—Less than an inch long, adults feed on the blood of deer, dogs, humans, and other animals. Infected females can transmit Lyme disease and other diseases to their hosts.

LARDER BEETLE (*Dermestes*) These small beetles can be serious pests wherever food is stored. The active, hairy larvae feed on all kinds of meat, leather, and other animal products. Four or more generations are produced annually. Widely distributed throughout the world.

HOUSE FLY—A common and despised fly associated with garbage and filth, these flies can reproduce a generation a month when temperatures are favorable.

GERMAN COCKROACH or croton bug, introduced from Europe, has become widely established in cities. Like the silverfish it is omnivorous and damages books. Its presence does not necessarily indicate uncleanliness, and it is not a proven carrier of disease like some flies.

SILVERFISH 0.5″

AMERICAN DOG TICK 0.2″

DEER TICK 0.3″

LARDER BEETLE 0.3″ and larva

HOUSE FLY 0.3″

GERMAN COCKROACH 0.6″

FOR MORE INFORMATION

The study of insects requires balancing what you learn first-hand with what you learn from others. It is best to supplement your collecting and field study with what you read. Books and museum exhibits can be very helpful. They reflect years of research and experience of experts.

MUSEUMS often have systematic, habitat, and local exhibits. Curators are glad to help you identify specimens. Check colleges and universities, as well as large city and state museums. Local inquiries are always best.

BOOKS TO READ The U.S. Department of Agriculture and many states publish bulletins on insects of economic importance. Write to Supt. of Documents, U.S. Govt. Printing Office, Washington, DC 20402, for price list of insect bulletins. Inquire locally for state publications. Some general books are listed below. Try them first before turning to more detailed or technical books.

Arnett, Ross H., and Richard Jacques, *Simon and Schuster's Guide to Insects,* 2nd edition, Simon & Schuster, New York, 1981.

Borror, Donald Joyce, Richard E. White, and Roger Tory Peterson, *A Field Guide to Insects: America North of Mexico* (Peterson Field Guides), Houghton Mifflin Co., Boston, 1998.

Chapman, R. F., *The Insects: Structure and Function,* Cambridge University Press, New York, 1999.

Milne, L., and M. Milne, *The Audubon Society Field Guide to North American Insects and Spiders,* Alfred A. Knopf, New York, 1980.

Waldbauer, G., *The Handy Bug Answer Book,* Visible Ink Press, Detroit, MI, 1998.

Web sites

The internet is full of interesting information about insects. Some keyboards to use in searches for these sites include insects, bugs, or entomology. You can also find information by searching for a particular insect such as praying mantis, beetle, butterfly, mosquito, or bee.

SCIENTIFIC NAMES

This list of the scientific names is included because the common names for insects sometimes differ from place to place or time to time. The list includes all the insects illustrated in this book. It follows the standard practice of listing the name of the genus first and the species second. A third name is a subspecies. If the species name is abbreviated, it is the same as the genus name that precedes it. The abbreviation "sp." means that the name applies to more than one species. The page number for each illustration is listed in bold type.

17 Mayfly: Ephemeroptera
Stonefly: Plecoptera
18 Anax junius
19 Ten-spot: Libellula pulchella
Blackwing: Calopteryx maculata
20 Diapheromera femorata
21 Mole: Neocurtilla hecadactyla
Camel: Ceuthophilus sp.
22 American: Schistocerca americana
Carolina: Dissosteira carolina
23 Lubber: Brachystola magna
Migratory: Melanoplus sanguinipes
24 Melanoplus femurrubrum
25 Gryllus pennsylvanicus
26 Bush: Scudderia sp.
True: Pterophylla camellifolia
27 Microcentrum sp.
28 Anabrus simplex
29 Periplaneta americana
30 Stagmomantis carolina
31 Mantis religiosa
32 Reticulitermes sp.
34 Forficula auricularia
35 Aphididae
36 Red-banded: Graphocephala coccinea
Lateral: Cuerna costalis
37 Potato: Empoasca fabae
3-banded: Erythroneura trincincta
Rose: typhlocyba rosae
38 Magicicada septendecim
39 Tibicen sp.
40 Stictocephala bisonia
41 Philaenus spumarius
42 San Jose: Quadraspidiatus perniciosus
Terrapin: Mesolecanium nigrofasciatum
43 Cottony: Icerya purchasi
Mealy Bug: Pseudococcidae sp.
Oystershell: Lepidosaphes ulmi
44 Harlequin: Murgantia histronica
Euchistus: euschistus sp.
45 Shield: Eurygaster alternata
Green Stink: Nezara viridula
46 Anasa tristis
47 Small: Lygaeus kalmii
Large: Oncopeltus fasciatus
48 Phymata fasciata
49 Chinch: Blissus leucopterus
Tarnished: Lygus lineolaris
50 Water Boatman: Corixa sp.
Backswimmer: Notonecta sp.
51 Water Strider: Gerris sp.
Giant Water: Lethocerus americanus
52 Head: Pediculus humanus capitis
Short-nosed: Haematopinus eurysternus
Crab: Phthiris pubis
Body: Pediculus humanus humanus
53 Corydalus cornutus
54 Golden-eye: Chrysopa aculata
Brown: Hemerobius sp.
55 Myrmeleontidea
56 Six-spotted: Cicindela sexguttata
Purple: C. purpurea
57 Calosoma scrutator
58 Carrion: Silpha americana
Garden: S. ramoso
59 American Burying: Nicrophorus amercanus
Hairy Burying: N. tomentosus

Hairy Rove: Creophilus maxillosus
60 Whirligig: Gyrindae
Diving: Dytiscus sp.
61 Water Scavenger: Hydrophilus triangularis
62 Lampyridae
64 Nine-spotted: Coccinella novemnotata
Convergent: Hippodamia convergens
65 Two-spotted: Adalia bipunctata
Fifteen-spotted: Anatis quindecimpunctata
66 Dung: Phanaeus vindex
Rhinoceros: Xyloryctes jamaicensis
Tumblebugs: Canthon laevis
67 Ox: Strategus capreolus
Eastern Hercules: Dynastes tityus
68 Macrodactylus subspinosus
69 Popillia japonica
70 Phyllophaga sp.
71 Cotinis nitida
72 Horn: Odontotaenius disjunctus
Stag: Pseudolucanus capreolus
73 Darkling: Carabidae sp.
Mealworm: Tribolium molitor
Tile-horned: Prionus imbricornis
74 Alaus oculatus
75 Locust: Megacyllene robiniae
Elder: Desmocerus palliatus
Pine Sawyer: Monochamus sp.
Flatheaded: Buprestis sp.
77 Epilachna varivestis
78 Colorado Potato: Leptinotarsa decemlineata
Striped Blister: Epicauta vittata
79 Spotted: Diabrotica undecimpunctata
Striped: Acalymma vittatum
81 Asparagus: Crioceris asparagi
Spotted: C. duodecimpunctata
82 Boll: Anthonomus grandis grandis
Plum: Cryptorhynchus nenuphar
Nut: Curculio sp.
Billbug: Sphenophorus sp.
83 Trichoptera
85 Papilio cresphontes
86 Danaus plexippus
87 Limenitis archippus
88 Banded: L. arthemis form proserpina
Red-spotted: L. arthemis astyanax
89 Junonia coenia
90 Great Spangled: Speyeria cybele
Meadow: Boloria bellona
Variegated: Euptoieta claudia
Regal: Speyeria idalia
91 Gulf: Agraulis vanillae
Silver-bordered: Boloria selene
Regal: Speyeria idalia
92 Speyeria cybele
93 Anicia: Euphydryas anicia
Baltimore: E. phaeton
94 Question Mark: Polygonia interrogationis
Comma: P. comma
95 Nymphalis antiopa
96 American Lady: Vanessa virginiensis
Red Admiral: V. atalanta
97 Vanessa cardui
98 Eyed Brown: Satyrodes eurydice
Pearly-Eye: Enodia anthedon
99 Large Wood: Cercyonis pegala
Little Wood: Megisto cymela
100 Gray: Strymon melinus
Purplish: Lycaena helloides
Little: L. phlaeas
Bronze: L. hyllus
101 Eastern: Everes comyntas
Marine: Leptotes marina
Spring: Celastrina argiolus
Pygmy Blue: Brephidium exile
102 Pieris rapae
103 Clouded: Colias philodice
Alfalfa: C. eurytheme
104 Spicebush: Papilio troilus
Parnassian: Parnassius phoebus smintheus
Giant: Papilio cresphontes
105 Zebra: Eurytides marcellus
Black: Papilio polyxenes
Tiger: P. glaucus
107 Spicebush: Papilio troilus
Zebra: Eurytides marcellus
Pipevine: Battus philenor
Black: P. polyxenes
Giant: P. cresphontes
Tiger: P. glaucus

SCIENTIFIC NAMES (continued)

108 Hobomok: Poanes hobomok
Silver-spotted: Epargyreus clarus
Southern Cloudy Wing: Thorybes bathyllus
109 Samia cynthia
110 Hyalophora cecropia
111 Callosamia promethea
112 Antheraea polyphemus
113 Automeris io
114 Actias luna
115 Eacles imperialis
116 Manduca quinquemaculata
117 Hyles lineata
118 Ultronia: Catocala ultronia
Locust: Euparthenos nubilis
119 Earworm: Helicoverpa zea
Borer: Ostrinia nubilalis
120 Salt Marsh Tiger: Estigmene acrea
Isabella: Pyrrharctia isabella
121 Orgyia leucostigma
122 Lymantria dispar
123 Webworm: Hypantria cunea
Tent: Malacosoma americanum
124 Alsophila pometaria
125 Synathedon exitiosa
126 Cydia pomonella
127 Thyridopteryx ephemeraeformis
128 Panorpa sp.
129 Tipula sp.
130 Anopheles: Anopheles sp.
Aedes: Aedes sp.
131 Culux sp.
132 Robber: Asilidae
Deer: Chrysops sp.
March: Bibionidae
Black Horse: Tabanus atratus
133 Tephritidae
134 Bluebottle: Calliphora sp.
Greenbottle: Phaenicia sp.
135 Syrphid: Toxomerus sp.
Tachinid: Tachinidae
Bee Fly: Bombyliidae
136 Tremex columba
137 Megarhyssa atrata
138 Potter: Eumeninae
Mason: Ancistrocerus birenemaculatus
139 Black-and-yellow Mud: Sceliphron caementarium
Cuckoo: Chrysididae
Blue Mud: Chalybion californicum
140 Oak Apple: Cynipidae
Elliptical Goldenrod: Gnorimoshema gallaesolidaginis
Goldenrod: Eurosta solidaginis
Blackberry: Diastrophus sp.
141 Camponotus sp.
142 Fire: Solenopsis geminata
Cornfield: Lasius niger
143 Little Black: Monomorium minimum
Pharaoh: M. pharaonis
Argentine: Iridomyrmex humilis
144 Velvet Ant: Dasymutilla sp.
Cow-killer: D. occidentalis
Thread-waisted Wasp: Sphex procerus
145 Sphecius speciosus
146 Polistinae
147 Bald-faced: Vespula maculata
Yellow Jacket: Vespinae
148 Bumble Bee: Bombus sp.
Sweat Bee: Halictinae
149 Leafcutting: Melittidae
Honey Bee: Apis mellifera
150 Buffalo Carpet: Anthrenus scrophulariae
Black Carpet: Attagenus megatoma
Clothes: Tinea pellionella
151 Dog Flea: Ctenocephalides canis
Sheep Ked: Melophagus ovinus
Duck Louse: Lipeurus squlidus
153 Silverfish: Lepisma saccharina
American Dog Tick: Dermacentor variabilis
Deer Tick: Ixodes scapularis
Larder: Dermestes lardarius
House Fly: Musca domestica
German Cockroach: Blatella germanica

INDEX

Asterisk (*) denotes pages on which illustrations appear.

MEASURING SCALE (IN 10THS OF AN INCH)
5
4
3
2
1